A HANDBOOK OF TOURISM AND HOSPITALITY EDUCATION: KEY INSIGHTS

A HANDBOOK OF TOURISM AND HOSPITALITY EDUCATION: KEY INSIGHTS

Prof. Dr. Sunil Kabia

SOCIETY Publishing

www.societypublishing.com

A Handbook of Tourism and Hospitality Education: Key Insights

Prof. Dr. Sunil Kabia

Society Publishing
224 Shoreacres Road
Burlington, ON L7L 2H2
Canada
www.societypublishing.com
Email: orders@arclereducation.com

ISBN: 978-1-77469-887-7 (Hardcover)

ABOUT THE AUTHOR

Prof. (Dr.) Sunil Kumar Kabia, B.Sc. (Hons), MTA, Ph.D., Dip IATA FIATA. He is currently working as Professor in Institute of Tourism & Hotel Management, Budelkhand University, India. He has 26 years of experience in teaching and handling other Academic positions. He is a member of board of studies and committees of various universities and governing council of various colleges. He has authored 4 books and published 35 research papers.

TABLE OF CONTENTS

List of Figures .... *xi*

List of Abbreviations .... *xiii*

Glossary .... *xvii*

Preface .... *xxi*

Chapter 1 Introduction to Tourism and Hospitality Education 1

1.1. Introduction 2

1.2. Types of Tourism 3

1.3. Hospitality Management Career Paths 5

1.4. History of Hospitality 6

1.5. History of Tourism 13

1.6. Experience is Career Ladder 21

1.7. Toward a New Paradigm 22

1.8. Scope and Future Will Not Disappear But Take Different Form 23

1.9. Future of Hospitality Education 24

1.10. Benefits of Personalized E-Learning Models for Tourism and Hospitality Education 24

1.11. Conclusion 27

References 28

Chapter 2 Tourism Management: Applications of Theories and Concepts of Tourism 29

2.1. Introduction 30

2.2. Concept and Functions of Management 31

2.3. The Concept of Tourism and its Definition 39

2.4. The Pillars of Tourism Theory 42

2.5. Tourism Theory Applications 46

2.6. General Systems Theory and Tourism 48

2.7. Conclusion 54
References 55

Chapter 3 Entrepreneurship Education in Tourism and Hospitality Management 57
3.1. Introduction to Entrepreneurship Education 58
3.2. Entrepreneur and Entrepreneurial Education 62
3.3. Characteristics of Entrepreneurship Education 65
3.4. Entrepreneurship Education and the New Role of Teacher 65
3.5. How Has Entrepreneurship Education Evolved? 69
3.6. Global Entrepreneurship Education and Research 70
3.7. Entrepretality 72
3.8. Preparing Entrepreneurs for Hospitality and Tourism Industry 74
3.9. Theoretical Grounds for Tourism Education and Entrepreneurship 77
3.10. Empirical Ground for Hospitality and Tourism Industry 79
3.11. Future Development of Hospitality and Tourism Education in Asia 80
3.12. Integration of Hotel Management Education and Innovative Entrepreneurship Education in Colleges 82
3.13. Conclusion 84
References 85

Chapter 4 Issues and Challenges in Tourism and Hospitality Education 87
4.1. Introduction 88
4.2. Sustainable Tourism 90
4.3. Tourism and Hospitality Issues Induced By Globalization 92
4.4. Developing Trends in Hospitality Education – New-Fangled Encounters 94
4.5. Getting Equipped the Future Dynamism of Hospitality Manpower 94
4.6. Present Scenario of Hospitality Education System 95
4.7. Factors Affecting Tourism Industry and Its Impacts on Global Economy of the World 96
4.8. Tourism Training and Education in Less-Developed Countries 97
4.9. Emerging Trends in Hospitality Education: New Challenges 101
4.10. Accommodation Sector 102
4.11. Conclusion 105
References 107

Chapter 5 Employability and Skills in Tourism and Hospitality 109

5.1. Introduction to Employability and Skills 110

5.2. Vocational Employability Skills 111

5.3. Graduate Employment 113

5.4. Types of Employment: SME, MNC, Non-Profit, Public, and Self 115

5.5. Employment in Tourism and Hospitality 117

5.6. Employment Opportunities in Tourism and Hospitality Sectors 119

5.7. Essential Skills for a Career in Tourism and Hospitality 122

5.8. Employment in the Tourism Industry to Grow Significantly 128

5.9. Challenges of Employments in Tourism and Hospitality 131

5.10. Future of Employment in the Hospitality Industry 133

5.11. Conclusion 137

References 138

Chapter 6 Mobilities, Tourism, and Travel Behavior 139

6.1. Introduction: Muslim Travel Behavior in Halal Tourism 140

6.2. Cultural Tourism and Event Tourism: Characteristics and Specificities 144

6.3. Theoretical Background and Hypothesis Development 146

6.4. What Is Long-Haul Travel? 152

6.5. Long-Haul Tourists 153

6.6. Travel Motivations 153

6.7. Analysis of the Spatial Range and Influencing Factors of Travel Behaviors 155

6.8. Cruise Traffic in the Mediterranean Area 157

6.9. Approaches and Procedures for Data Achievement 160

6.10. Dynamics of the Colombian Medical Tourism 162

6.11. Systemic Competitiveness of Associative Character: A Theoretical Approach 164

6.12. Conclusion 166

References 168

Chapter 7 Contemporary Developments and Perspectives in International Health Security 171

7.1. Introduction 172

7.2. Expanding the Definition and the Scope of Health Security 173

7.3. Emerging Infectious Diseases 174

7.4. Chronic Health Conditions and Access to Care.....175
7.5. Trauma Systems.....177
7.6. Emergency Medical Services.....179
7.7. What Is Ransomware?.....182
7.8. International Health Security: Potentially Harmful Aspects of Social Media Use.....185
7.9. Human Trafficking.....188
7.10. Theories of Morality.....189
7.11. Health and Safety at Work.....194
7.12. Advantages of Microfluidic Lab-on-a-Chip Technology.....196
7.13. Conclusion.....198
References.....199

Chapter 8 Tourism: From Empirical Research to Practical Application.....201
8.1. Community Participation Toward Tourism Development.....202
8.2. Types of Community Participation.....205
8.3. Factors Influencing Community Participation.....206
8.4. Community Development Strategies for Tourism Development in Langkawi Islands, Malaysia.....207
8.5. Ecotourism and its Role in Sustainable Development.....211
8.6. Tourism, Competitiveness, and Economic Growth: A New Analytical Model.....218
8.7. Interaction Between Cultural/ Creative Tourism and Tourism/Cultural Heritage Industries.....223
8.8. Conclusion.....229
References.....230

Index.....231

LIST OF FIGURES

Figure 1.1. Tourism distribution chain

Figure 1.2. Global Hospitality education expo

Figure 2.1. Henry Fayol

Figure 2.2. Giza Pyramids

Figure 2.3. Ibn Battuta Portrait

Figure 2.4. Marco Polo traveling

Figure 2.5. Hajj

Figure 2.6. Aristotle

Figure 3.1. Reshaping the World through Entrepreneurship, Education and Employment: Klaus Schwab, Park Geun-Hye

Figure 3.2. Education Development Centre, Enterprise Way, Spennymoor

Figure 3.3. YSEALI Cambodia – eco-tourism and entrepreneurship for all

Figure 3.4. XLRI Campus

Figure 3.5. Entrepreneurship Development Institute of India (EDI)

Figure 3.6. Expectancy theory of motivation (ETM)

Figure 4.1. Randy Durband, CEO Global Sustainable Tourism Council

Figure 4.2. Call Center

Figure 4.3. Select-service hotels

Figure 4.4. Full-Service Hotel

Figure 5.1. Women in vocational education

Figure 5.2. Interpersonal skills between two person

Figure 5.3. Panel discussion on the organization of work and production, The Future of Work We Want: A Global Dialog

Figure 5.4. Working in cruise

Figure 5.5. Teamwork Training

Figure 5.6. Two employees sat together working on problem solving skills. One is using a laptop while the other speaks, and they are surrounded by puzzle pieces

Figure 5.7. Multi-Tasking Efficiency Manager

Figure 5.8. The United Nations World Tourism Organization (UNWTO)

Figure 6.1. Halal friendly destination

Figure 6.2. Islamic Centre

Figure 6.3. Pilgrimages and Santiago Museum

Figure 6.4. Parachute adventure

Figure 6.5. West Mediterranean sea areas

Figure 6.6. Doctor with stethoscope and globe in his hand. Medical network and health care in Europe and Africa, concept of world medicine, epidemic control, medical tourism

Figure 7.1. Infectious disease

Figure 7.2. Chronic health and access to care

Figure 7.3. Emergency medical services

Figure 7.4. Human trafficking

Figure 8.1. 40th Edition of UNESCO World Heritage Session

Figure 8.2. The Langkawi Development Authority (LADA)

Figure 8.3. Ecotourism: Tourists waiting to take photographs of orangutans

Figure 8.4. Xiaodubai Rural Tourism Area scene

Figure 8.5. Porter's Diamond model

Figure 8.6. Kernavė Archaeological Site (Cultural Reserve of Kernavė) (Lithuania)

LIST OF ABBREVIATIONS

AACSB	American Assembly of College Schools of Business
AI	Artificial Intelligence
AICTE	All India Council for Technical Education
ATLAS	Association for Tourism and Leisure Education
ATLS	Advanced Trauma Life Support
BPO	Business Process Outsourcing
CAS	Council for the Advancement of Standards
CBE	Community-Based Entrepreneurship
CBSE	Central Board for Secondary Education
CHC	Chronic Health Conditions
COVID-19	Coronavirus Disease 2019
ED	Emergency Departments
EDI	Entrepreneurship Development Institute of India
EID	Emerging Infectious Diseases
EMS	Emergency Medical Services
EMT	Emergency Medical Techniques
EOL	End of Life
ETM	Expectancy Theory of Motivation
FEE	Foreign Exchange Earnings/Foundation for Environmental Education
FHRAI	Federation of Hotel & Restaurant Associations
FTE	Full Time Equivalent
GDP	Gross Domestic Product
HIR	High Income Region

HIS	International Health Security
HIV	Human Immunodeficiency Virus
HR	Human Resource
HRD	Human Resource Development
IGNOU	Indira Gandhi National Open University
IHM	Indian Hospitality Management
IHRA	International Hotel and Restaurant Association
IIM	Indian Institute of Technology
IIT	Indian Institute of Management
ILO	International Labor Organization
IP	Internet Protocol
IT	Information Technology
ITTM	International Technology Transfer Management
IUF	International Union of Food
LADA	Langkawi Development Authority
LDC	Less Developed Countries
LMIR	Lower Middle Income Region
MBA	Masters in Business Administration
MFT	Master File Table
MNC	Multinational Corporation
NCHM	National Council for Hotel Management
NCHMCT	National Council for Hotel Management and Catering Technology
NGO	Non-Governmental Organization
OECD	Organization Economic Co-operation and Development
OP	Original Position
OS	Operating System
PMS	Premenstrual syndrome
POC	Point-of-Care

QAA	Quality Assurance Agency
RAAS	Ransomware as a Service
RAMS	Risk Amplification through Media Spread
RevPAR	Revenue per Available Room
SARS	Severe Acquired Respiratory Syndrome
SEO	Search Engine Optimization
SIWES	Students Industrial Work Experience
SM	Social Media
SME	Small Medium Enterprises
STD	Sexually Transmitted Disease
TE	Teacher Educators
TEQSA	Tertiary Education Quality and Standards Agency
THSC	Tourism and Hospitality Sector Council
TLG	Tourism Led Growth
TVET	Technical and Vocational Education and Training
UNDP	United Nations Development Programme
UNESCO	United Nations Educational, Scientific, and Cultural Organization
UNWTO	United Nations World Tourism Organization
VET	Vocational Education and Training
VPN	Virtual Private Network
WHS	World Heritage Site
WTO	World Tourism Organization

GLOSSARY

A

Accommodation – The term accommodation refers to a room, building or lodging which provides shelter for a person to stay, sleep and live.

Adriatic – relating to the region comprising the Adriatic Sea and its coasts and islands.

B

Bio politics – Bio politics is a critical term used with some variation across the fields of political theory, international relations, cultural studies, critical sociology, and globalization studies.

Biodiversity – Biodiversity or biological diversity is the variety and variability of life on Earth.

Biologist – an expert in or student of the branch of science concerning living organisms.

Bioterrorism – Bioterrorism refers to the intentional release of biological agents or toxins for the purpose of harming or killing humans, animals or plants with the intent to intimidate or coerce a government or civilian population to further political or social objectives.

C

Cognitive – Relating to, being, or involving conscious intellectual activity

Concierges – A usually multilingual hotel staff member who handles luggage and mail, makes reservations, and arranges tours.

Conservation – Conservation is a plan for avoiding the unnecessary use of natural materials such as wood, water, or fuel

Consortium – A consortium is an association of two or more individuals, companies, organizations or governments with the objective of participating in a common activity or pooling their resources for achieving a common goal.

Cultural Erosion – The process of a civilization, group or organization losing its core elements — often because of new elements taking their place.

Cultural Heritage – Cultural heritage is the heritage of tangible and intangible heritage assets of a group or society that is inherited from past generations.

E

Ecotourism – Ecotourism is a form of tourism involving responsible travel to natural areas, conserving the environment, and improving the well-being of the local people.

E-Learning – A learning system based on formalized teaching but with the help of electronic resources is known as E-learning.

Employable – Someone who is employable has skills or abilities that are likely to make someone want to give them a job.

Employment – Employment most generally means the state of having a paid job.

Enterprise – Enterprise is another word for a for-profit business or company, but it is most often associated with entrepreneurial ventures. People who have entrepreneurial success are often referred to as "enterprising."

Entrepreneurship – Entrepreneurship is the ability and readiness to develop, organize and run a business enterprise, along with any of its uncertainties in order to make a profit.

Entrepretality – Entrepretality is a new term that we created that describes the objective of entrepreneurship education in the hospitality industry.

G

Globalization – Globalization is the word used to describe the growing interdependence of the world's economies, cultures, and populations, brought about by cross-border trade in goods and services, technology, and flows of investment, people, and information.

H

Halal Tourism – Halal tourism is a subcategory of tourism which is geared towards Muslim families who abide by rules of Islam.

Hedonic – Relating to, or characterized by pleasure

Hospitality – Generous and friendly treatment of visitors and guests or hospitable treatment.

Human Resource Development – Human resource development (HRD) is defined as the cultivation of an organization's employees.

Human Trafficking – Human Trafficking is the recruitment, transportation, transfer, harboring or receipt of people through force, fraud or deception, with the aim of exploiting them for profit.

Hypothesis – An idea or explanation for something that is based on known facts but has not yet been proved

I

Infrastructure – Infrastructure is defined as the basic physical systems of a business, region, or nation and often involves the production of public goods or production processes.

M

Management – Management is a process that is very important for day-to-day life and it is used in household-to-business organizations to ease the work process.

Manifestation – A sign of something existing or happening

Mediterranean – Characteristic of the Mediterranean Sea, the countries bordering it, or their inhabitants.

Morbidity – Morbidity is any condition that isn't healthy. It can refer to mental or physical illness. A person with a morbidity may not live as long as someone who is healthy.

Multisensory – Involving or using more than one of the senses.

P

Pandemics – Pandemics are large-scale outbreaks of infectious disease that can greatly increase morbidity and mortality over a wide geographic area and cause significant economic, social, and political disruption.

Phenomenon – A fact or situation that is observed to exist or happen, especially one whose cause or explanation is in question.

Philosophically – In a way that calmly accepts a difficult situation

Pilgrimages – The course of life on earth

Pre-Hospital – Pre-hospital care is provided by emergency medical services (EMS) responders, who are the initial health care providers at the scene of disaster.

Professionalism – A professional is a member of a profession or any person who works in a specified professional activity.

Psychologist – A psychologist is someone who studies the mind and behavior.

S

Salarium – Salary comes from the Latin word salarium, which also means "salary" and has the root sal, or "salt."

Staffing – Staffing refers to the continuous process of finding, selecting evaluating and developing a working relationship with current or future employees.

Sustainable Development – Sustainable development is development that meets the needs of the present without compromising the ability of future generations to meet their own needs.

T

Tourism – The business of providing services such as transport, places to stay, or entertainment for people who are on holiday.

V

Virtual – Being such in essence or effect though not formally recognized or admitted.

Vocational – Relating to the special skills, training, etc., that you need for a particular job or occupation.

PREFACE

In an increasingly globalized world and the shifting paradigm of urban living, the demand for hospitality and tourism has grown significantly throughout the world. It has become crucial to offer competency-based vocational education in this constantly growing industry in order to give students knowledge of the front desk, the significance of grooming, effective hotel communication, the use of ICT tools, and other desirable qualities of front office staff, among other topics.

The book provides an introduction and overview of the worldwide hospitality and tourism sector, as well as information on its growth and development, different industry segments, and the features that define them, trends, and pressing issues.

The hospitality industry is a billion-dollar industry and has great future prospects and increasing demand because of the global mobility of people for various purposes, such as tourism, business, events, and sports. Hospitality is one of the most interesting and challenging industries as it involves being sensitive towards the need and desires of guests coming from different backgrounds and for different purposes.

On the other hand, tourism is much more than just an industry; it also influences social, cultural, political, and environmental development. The development of communication, transportation, lodging, and other consumer-related services has all benefited from the tourism industry's emergence as a global economic and social force that has the biggest multiplier effect. With the advent of several technological and economic advancements as well as an ever-increasing concern for the environment, tourism has undergone a significant transformation recently.

The main objective of this book is to present a perceptive and reliable assessment of the numerous difficulties influencing the fields of tourism, hospitality, and events education, as well as to showcase the original, inventive, and cutting-edge ways that educators are addressing these concerns.

The book explains in detail the various types of tourism, distribution chain, hospitality management skills, history, challenges, growth, and development, along with changing trends in the tourism and hospitality industry.

It also discusses in detail entrepreneurship education in tourism and hospitality management, the characteristics of entrepreneurship education, and explains how to prepare entrepreneurs for the hospitality and tourism industry. It also focuses on the future development of hospitality and tourism education in India and the integration of hotel management education and innovative entrepreneurship education in colleges.

Issues and challenges in tourism and hospitality education have also been highlighted. It also brings forth employment opportunities with all the essential skills needed for a career in the tourism and hospitality industry.

A detailed understanding of International Health Security and different health challenges has been addressed in the tourism sector. Towards the end, Ecotourism has been highlighted along with the benefit of Cultural Tourism.

Practicing professionals from the field of Hospitality and Tourism comprised the team of authors for this book. We thankfully acknowledge their contribution to completing the book in record time. We express our gratitude to the experts who meticulously proofread the entire book and offered us insightful criticism that resulted in appreciable changes. I hope this book will serve as a useful resource on this subject.

Chapter

1 INTRODUCTION TO TOURISM AND HOSPITALITY EDUCATION

Contents

1.1. Introduction 2

1.2. Types of Tourism 3

1.3. Hospitality Management Career Paths 5

1.4. History of Hospitality 6

1.5. History of Tourism 13

1.6. Experience is Career Ladder 21

1.7. Toward a New Paradigm 22

1.8. Scope and Future Will Not Disappear But Take Different Form 23

1.9. Future of Hospitality Education 24

1.10. Benefits of Personalized E-Learning Models for Tourism and Hospitality Education 24

1.11. Conclusion 27

References 28

This chapter will cover every aspect of tourism and hospitality education. This chapter also explains the types of tourism and hospitality management career paths. This chapter also provides highlights on the history of tourism and hospitality. This chapter also helps to understand the role of hospitality education. This chapter also explains the challenges and opportunities in hospitality education. At the end of this chapter, this chapter provides highlights on the future of hospitality education and the benefits of personalized e-learning models for tourism and hospitality education.

1.1. INTRODUCTION

Hospitality is one of the oldest industries, as far back as the innkeepers and taverns of biblical times. On the other hand, tourism is a more recent innovation that had its start in Europe, with Switzerland being one of the first nations to provide specialized lodging and services for visitors.

The word "hospitality" is derived from the Latin word "hospes," which means visitor or stranger, and signifies to welcome visitors or to provide a home away from home. The hospitality and tourism sector is vast sector that encompasses all commercial pursuits that either directly or indirectly support or are reliant upon travel and tourism. This industry includes sectors like Hotels & Resorts, Restaurants & Catering, Night Clubs & Bars, etc.

The tourism and hospitality education sector is part of the larger tourism and hospitality industry and is committed to supplying the sector with a workforce that is committed, properly educated, and adequately trained.

Tourism is a dynamic and competitive industry that requires the ability to adapt continuously to consumers' changing wants and desires. It is now acknowledged that tourism is a significant global economic activity.

With the increase in its significance, governments, companies in the public and commercial sectors, and academic institutions have all begun to pay more attention. The industry of tourism and hospitality education is complicated and has many facets, and this book serves as an introduction to it.

It is primarily intended for two groups of readers: tourism and hospitality management students and professionals in the field who wish to learn more about the organization, constituent tasks, and working environment in which they are employed. The aim of the book is to give a thorough overview of the tourism sector in order to promote future research and stimulate interest in the field.

Tourism refers to the movement of people between locations and nations outside of their normal environments. This movement may occur for a variety of reasons, including amusement, necessities for the individual, or for business, professional, or personal reasons. In fact, it can be viewed as a social, economic, and cultural phenomenon that involves the flow of information, ideas, cultures, languages, foods, and other things.

Tourism is essential for the success of many economies throughout the world. It significantly contributes to the growth of any nation's economy, by generating jobs, assisting in the development of infrastructure, and fostering a sense of cultural interchange between locals and visitors.

For many countries in the world tourism is the only source of income. The jobs generated by tourism are distributed across several service-related industries, but they are also indirectly generated in the agricultural, communication, health, and educational fields.

Tourism is a great opportunity for foreigners to learn about a new culture, but it also creates many opportunities for local citizens. Many tourists visit new places to learn about the local cultures, rural life, traditions, etc., which benefits small business owners, restaurants, and shops in the area.

Countries whose economies are mostly based on tourism make significant investments in enhancing their infrastructural projects, such as motorways, new roads, flyovers, and tunnels, which have a favorable effect on the country's development and benefit its population. Improved infrastructure promotes the efficient and secure flow of products and services, which in turn promotes economic growth.

Since tourism is a primary driver of new airport construction, which has helped to increase global connectivity, it has also had a significant impact on transportation infrastructure around the world. This has aided in the exposure of local businesses by giving them access to the international market. Tourism initiates cultural and traditional exchanges between visitors and locals, and over time, it has the potential to significantly alter national lifestyles, dietary preferences, and cultural practices. Events like exhibitions, conferences, and other gatherings for foreigners are profitable for the host country.

1.2. TYPES OF TOURISM

Tourism can be majorly categorized into three groups; domestic tourism, inbound tourism, and outbound tourism.

1.2.1. Domestic Tourism

It refers to a visitor's actions both inside and outside the borders of the country of their residency.

1.2.2. Inbound Tourism

It refers to the acts of a tourist from outside their country of residence.

1.2.3. Outbound Tourism

It refers to a resident visitor's activities while they are traveling outside of their country of residence.

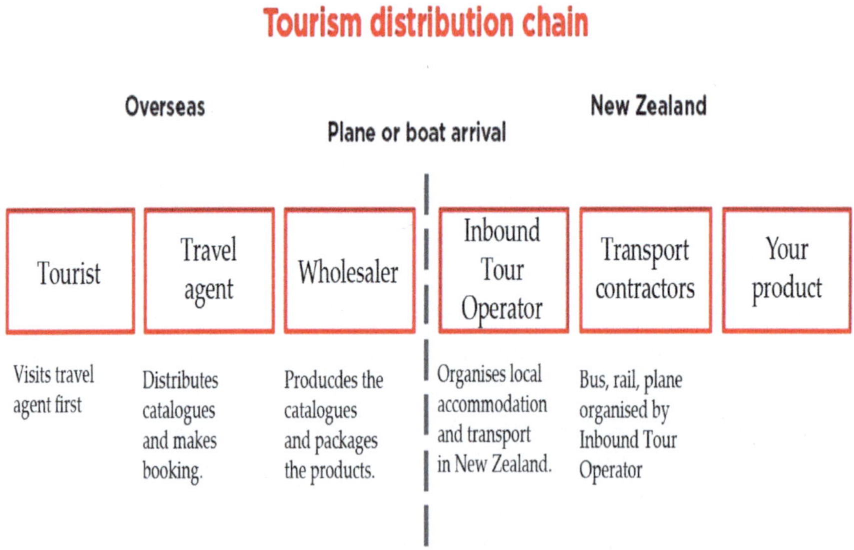

Figure 1.1. Tourism distribution chain.

Source: Image by Wikimedia Commons

According on the type of activity a visitor engages in while traveling, tourism can be further divided into categories such as active tourism, ecotourism, food tourism, wellness tourism, health tourism, cultural tourism, historical tourism, and recreational tourism, etc. (Figure 1.1).

Since hospitality education has become a driving force for emerging countries, these economies have identified hospitality as a means of achieving their desired levels of development and progress. Education in hospitality and tourism has long been considered important for the growth of tourism in any area.

Education in the hospitality and tourism industry is crucial for producing the correct kind of workforce, which can then improve planning and add the necessary professionalism to the sector. Hospitality education has a number of problems, such as a lack of standardization and uniformity in the curriculum, which causes problems for both students and employers.

Management of the daily administrative, operational, and commercial operations of companies in the hospitality sector falls under the broad category of hospitality management. The word "hospitality management" is more general and refers to the management of people in a variety of industries, including nightclubs, casinos, restaurants, transportation, and even hotels. Maintenance and housekeeping, spa services, concierge services, reception, and many more areas may fall under the purview of a hospitality manager.

Depending on the environment in which they work, hospitality managers are assigned different duties.

Chiefly, these can include the following:

- Accounting and budgeting
- Interviewing, training and managing staff
- Assisting with daily operations
- Optimizing business performance
- Fielding customer complaints and inquiries
- Ensuring health and safety compliance

1.3. HOSPITALITY MANAGEMENT CAREER PATHS

Traditional hospitality jobs frequently begin in hotels and resorts, and the lodging sector is diversifying as more new properties and innovative ideas enter the market. Not everyone is suited for a career in hospitality management, which may be demanding yet rewarding.

There are many different sorts of employers that provide employment prospects in hospitality management, including hotel chains, conference and exhibition centers, event venues, bars and pubs, restaurants and fast-food joints, airlines, as well as the public sector.

Alongside the roles listed above, hospitality management skills may open the door to a career in a wide range of different roles, including:

- Marketing/Sales
- Public relations
- Human resources
- Customer services
- Tourism
- Finance/Accounting

1.4. HISTORY OF HOSPITALITY

The term "hospitality" initially appeared in use in the 14th century, before the term "tourist." It is derived from the Latin word hospes, which includes the phrases foreigner, host, and guest (Latdict, 2014).

The origins of hospitality and tourism may be traced back to the 1700s, when young, affluent Europeans started taking tours throughout the continent in order to widen their horizons and acquire an education in history, the arts, and cultures.

The development of tourism and hospitality degree programs went through stages: Slow progress was made in the 1970s, followed by increased progress in the 1980s, and major advancement from the 1990s to the present (Breakey and Smith, 2008). Such stages of development varied depending on the nation, the national educational system, and the structure of the tourism business.

Through a process of academic and professional progression in response to both industry and student demand, tourism education has evolved in many nations from training created to satisfy the skill demands of the local hotel and restaurant businesses (Baum,2005).

Learning and exploration are at the core of universities' missions. Academics work in the fields of education, research, and community service. An academic department must function within the parameters of the university's resource allocation policies as well as the international research, educational, and business environments (Ringwood et al., 2005; Coates and Goedgebuure, 2012).

The issue facing tourism and hospitality-related programs at universities is that they must uphold established academic standards to align with other business-related bachelor degree programs while also taking the demands of the industry into account as they train future employees and managers (Harper et al., 2005; Raybould and Wilkins, 2005; Dawson, 2014).

Future students' decision to enroll in a program at a university is significantly influenced by their employability (Jauhari, 2013). Political and educational organizations have put pressure on higher education institutions to help students become employed. These institutions are expected to adopt all techniques and tactics that increase students' prospects of finding employment (Harvey, 2010).

The aim of education in the tourist and hospitality sectors is to provide the industry with competent graduates who are able to manage business-related challenges professionally (Enz et al., 1993). There would be conflict between academics and industry affiliates over the courses taught to students if the curriculum was primarily established from an academic standpoint while ignoring the industry's dynamic environment (Zhang et al., 2001; Li and Li, 2013).

The agreement between parties and an educational institution regarding the educational experiences of students is referred to as the curriculum (Tsui, 2009). In order to prevent discrepancies in expectations and perceptions on both sides, the gap between academia and the industry needs to be closed (Harkison et al., 2011).

Rules and regulations placed on universities make it difficult for them to adapt to the rapid changes in the tourism environment. It is a lengthy and time-consuming procedure, but it is worthwhile to make even the simplest changes to a study plan, which would involve proposals and committee reviews (Quain et al., 2014). The curriculum reform cycle includes several connected components, such as needs analysis, goal planning, program implementation, and evaluation of program (Kirkgoz, 2009).

During the 1980s and the early nineties, the majority of program in the United States were focused on operating skills. There has been a substantial shift in the last ten years, with university programs in hospitality and tourist management differentiating themselves from technical institutions by include operational and management-related courses in their curriculum. By using this strategy, graduates will be seen as being more desirable by potential employers (Rappole, 2000).

Educational institutions should focus on enlisting students who want to finish their degrees and commit to a career in the sector. A high attrition rate makes it difficult for the educational institution to fulfill its objective (Astin, 1993; Rudd et al., 2014).

According to Tinto (1982) and Martnez et al. (2015), the problem known as undergraduate student attrition is brought on by students who enroll but drop out before finishing their course. Less than half of graduates enter the tourism business, despite having received a higher professional education that is meant to prepare them for such a career, according to research in the same vein (King et al., 2003).

The choice of students' careers in tourism and hospitality, as well as their intention to work in the field once they graduate, is a significant problem. On one hand, despite the fact that tourism management programs are designed to prepare students for careers in the sector, they may not always opt to work in the field after they graduate.

On the other hand, even if tourism was their initial career choice, graduates may not stay in the field for the long term, leading to a greater attrition rate than usual. The repercussions are negative; and the industry may soon experience a significant shortage of competent workers, especially at the management level (King et al., 2003; Chuang and Dellmann-Jenkins, 2010; Lo et al., 2014).

Several factors, including internship experience and familial influence, have an impact on how committed students and graduates are to pursuing careers in the field (Atef and Al-Balushi, 2017a). According to researchers who looked at the perceptions of undergraduate tourism and hotel management students, students without any prior job experience had neither a favorable nor a negative opinion of professions in tourism and hospitality.

More significantly, students who participated in internship programs and other related work experiences in college had negative perceptions when they encountered irregular working hours, lacked job security, received low pay, and lacked development opportunities (Casado, 1992; Roney and ztin, 2007; Lu and Adler, 2009). Further research indicates, good internship methods increase interns' desire to pursue employment in the hotel industry after graduation (Busby, 2003; Jauhari and Manaktola, 2006; Chen and Shen, 2012).

Some academics were concerned about measuring the impact of families and parents on people's career paths. Parents and family (combined) are the most influencing factors in high school students' career decisions, but friends, teachers, and counselors played a much smaller role in forming student attitudes and career choices.

Young people form many of their attitudes about work and career from interactions with the people around them (Cothran and Combrink, 1999). Family engagement and impact can however vary, bringing both positive and negative effects. Youth's incapacity to make decisions may be influenced by parents' lack of involvement in the process of choosing a vocation (Lankard, 1995; Haven and Botterill, 2005; Ferry, 2006).

Growing focus is being paid to non-traditional graduate employment, particularly in relation to entrepreneurship and small and medium-sized organizations (SMEs) (Connor and Shaw, 2008; Hart and Barratt, 2009; Wilton, 2011; Knight and Mantz, 2013; Dias et al., 2016a, 2016b). However, it has been asserted that employment with large firms is preferred by hospitality students above employment with SMEs (Jenkins, 2001; Pittaway and Thedham, 2005; Nachmias et al., 2014).

The study conducted by Atef and Al-Balushi (2015) to gauge entrepreneurial intentions among students at Sultan Qaboos University (College of Arts and Social Sciences), located in the Sultanate of Oman, revealed that due to the condensed workweeks and significant benefits associated, respondents prefer to enter the labor market in the government (public) sector rather than the private sector, and only a small number of them pursue entrepreneurship (self-employment).

King et al. (2003) discovered that up to half of all graduates will never work in the business or leave after a short time in their comparative research of hospitality and tourism graduates in Australia and Hong Kong. For supervisors of academic programs and the available study plans, this reality poses a serious problem.

Should study plans prioritize specialized tourism and hospitality courses, knowing that half of the graduates may not find these courses helpful for their jobs, or should the priority be on more general business courses that could help graduates looking for employment in the sector or elsewhere?

There is a difference between tourism education programs that take a general approach (education about tourism) and those that focus on operational aspects of the sector (education for tourism) (Stuart-Hoyle, 2003).

Researchers contend that studies should be about tourism rather than for tourism, curriculum need to have a wider viewpoint, and the vocational emphasis should not take precedence (Ryan, 1995; Gilbert and Guerrier, 1997; Chen and Groves, 1999; Stuart, 2002; Shaikh and Syed, 2013). According to Biggs and Tang (2007), the emphasis should be shifted from a conventional perspective on curriculum to a more creative one that can keep up with the industry's growing rhythm. In the past, jobs in the hospitality sector did not necessitate a formal degree. Even in 2022, many entry-level, management, and supervisory positions can be filled without a college degree. In fact, a number of the top CEOs in the hotel, cruise, airline, travel agency, and restaurant industries began their careers in entry-level positions right out of high school before working their way up through a combination of grit, perseverance, and on-the-job training to become the leaders of their companies.

However, things have changed. Globally, more positions in the tourism industry are being added that call for graduate degrees, college degrees, or equivalent training. Hospitality and tourist firms may now require a minimum of a bachelor degree to enter their manager in training roles in fields including data analytics, artificial intelligence, and forensic accounting.

1.4.1. Early Education

The first hospitality and tourism college-level programs were developed in Europe, and they were primarily concerned with etiquette, guest service, a working understanding of food and drink, and—to a lesser extent—business management. The Cornell program started in the United States in 1922.

This first degree program in the United States was started by hotel tycoons from the American Hotel Association (now the American Hotel & Lodging Association). It was the first of its kind; they even gave it the term "hospitality management" (Going forward, the term "hospitality" will refer to the vast industry of travel, tourism, hospitality management, gastronomy, etc.).

The hospitality sector has evolved over time to its current form, which includes multinational corporations in every specialized sector, including travel agencies, meeting and event planners, destination marketing firms, hotels and resorts, casinos, cruise lines, airlines, car rental agencies, and so on.

A sophisticated level of software, hardware, standard operating procedures, accounting, and finance is increasingly required even for entrepreneurial. Currently, all types of hospitality enterprises need business expertise in every field, from accounting and finance to information technology, from economists to digital marketers, and from supply chain managers to more traditional operations specialists handling the day-to-day operations.

College-level hospitality programs have flourished as our sector has become more professional and has grown to demand more education from its employees. The number of programs has soared in the century from 1922 (when Cornell made its debut in the United States) to 2022.

There are currently hundreds of degree programs with enormous variety and specialized specialty areas, according to the "Guide to College Programs in Hospitality, Tourism, & Culinary Arts (2021)" published by the International Council on Hotel, Restaurant and Institutional Education. These include two-year (associate), four-year (baccalaureate), graduate (Master of Science, MBA, or comparable), and for-credit and non-credit certificates, as well as professional certifications of all kinds.

Between 1980 and 2010, programs alone in the United States had particularly notable increase before a plateau appeared. National enrollments have been declining for the previous 12 years, particularly in the higher education field of culinary arts. Numerous programs have recently developed "niche" educational techniques in an effort to reverse this loss.

A few specialized degree programs include hospitality leadership, sustainability in hospitality operations, and hospitality entrepreneurship. Additionally, several programs have now relocated to the colleges of business at their individual institutions, while others have chosen to become independent programs or correlate with other fields like nutrition, recreation and parks, education, or other fields.

There are much more instances of Hospitality College programs transferring into business schools than there are of those leaving business schools across the nation. In fact, when Cornell officially moved into the SC Johnson College of Business as the Nolan School of Hotel Administration, it cemented the idea that studying hospitality management in business schools is a valid, appropriate, and contemporary style of instruction.

In fact, a number of brand-new specialty markets all fit within the traditional business functions with a focus on hospitality. It is probably best if a combination of IT specialists, engineering experts, marketing experts, and hospitality experts provide courses on utilizing artificial intelligence to secure robotics that specialize in maximizing worker efficiency in the hospitality industry. We would rely on people with accounting degrees, potentially MBAs, who are also familiar with the hospitality sector if we wanted to hire forensic accountants for our hotel management company. These combinations of learning, specialization, and future career planning have excellent synergy in the business school environment.

1.4.2. Changing Styles

The range of educational approaches offered opportunities to keep and/or grow enrollments. It represented a historical departure from the training of a future general manager as "figurehead," etiquette leader, and culture creator. These same abilities are still required, but they must be acquired along the way through more conventional business schooling.

COVID had an impact on the hospitality sector's workers, owners, and those who educate the sector's future leaders as the world headed into late 2019 or early 2020. Although the rate of mass resignations from the hotel industry has decreased, labor issues still exist.

According to Florida Atlantic University research, nearly 70% of workers in the hotel industry planned to change jobs by 2020. The Bureau of Labor Statistics reported the food service industry's quit rate to be 6.8% in April 2022, an all-time high when compared to the industry's average of roughly 2.9%. In actuality, hotels still employ between 10% and 20% fewer people than they require.

The majority of those who were dislocated during COVID are talking to their neighbors, loved ones, and kids. Many parents are advising their kids to consider majoring in fields other than hospitality. The sector has boosted entry-level tariffs to previously unheard-of heights in an effort to preserve business operations and the enormous demand from tourists. In the short term, this was a survival strategy; the effects on profitability are still to be seen. However, the labor situation has not entirely improved.

With greatly increased entry-level wages, one might assume that students would be more motivated than ever to major in hospitality. Higher pay rates and more vacant employment than ever before. This demand among students has not yet been met. Anecdotally, regular talks with deans and directors of hotel management college programs show enrollment declines from 5% in the best cases to over 50% among the severely hit programs.

1.4.3. Combination Efforts

Where do we proceed from here, then? Having worked in the industry for more than 25 years and spent an additional 15 years teaching full-time in the hospitality sector, I believe that the future should be approached with caution. Through automation, robots, and the use of artificial intelligence, the sector will continue to eliminate full-time equivalent positions, resulting in a decrease in the total number of people worldwide. The demand for people with higher levels of education and critical thinking abilities will rise.

Additionally, following the route we have previously charted, both independent owner/operators and major, international chains and brands will favor the business-school model of hospitality education. However, the value of good customer service, manners, and conventional hospitality skills will always be present. It requires a coordinated effort.

The essential hospitality values and teachings must be maintained by schools that adopt a business school model, while the traditional business discipline lessons must be improved and expanded in programs that remain independent.

Future employees will demand greater possibilities for remote work. Once they acquire a taste for any potential distant roles, fewer will be able to endure the constant stress that is frequently placed upon them by demanding visitors. Our workforce will continue to assess the advantages, fulfilment, and experiences that hospitality businesses provide or do not supply. The overall business basis is thus advantageous to many. They have a variety of options for their future job path. Awakening, reconsideration, and a substantial shifting of the ground beneath our feet are currently occurring on all fronts, including those related to operations, career paths, income, and education. According to an instructor in Florida, one of the top tourist destinations on earth, there will always be a demand for hospitality employees and those programs and places that educate prospective workers. However, if I thought enrollments in the hospitality and tourism industries will be as robust as they were from 1980 to 2010, I would prove too much of an optimist. In fact, a lot of people have left the sector, and a lot more will keep looking elsewhere. It is a plain and simple fact. When automation is taken into account, the number of people who would like to obtain a college degree in hospitality management becomes more realistic.

Only the most effective programs that offer employers value will be kept after the global decrease in FTEs. As automation, robotics, and artificial intelligence (AI) become more and more integrated into everything we do in the hospitality workplace, our next step may be to form collaborations with engineering and IT. Additionally, all industries—including the hospitality sector—will see continued advancements in social media, digital marketing, search engine optimization (SEO), the creation of new apps, and related technologies.

1.5. HISTORY OF TOURISM

International travel was one of the most significant industries in the world at the beginning of the twenty-first century, and its effects could be felt from the Arctic to the Antarctic. Therefore, the history of tourism is quite interesting and significant. The history goes back much further than the 18th-century invention of the term "tourist."

Ancient Greece and Rome can lay claim to being the birthplaces of both "heritage tourism" (dedicated to the celebration and appreciation of historic sites of acknowledged cultural importance) and beach resorts. Organized travel with supporting infrastructure, sightseeing, and an emphasis on essential destinations and experiences can be found there.

In Mesopotamia and India, 2000 years before Christ Since the dawn of civilization, travel for trade has been a significant aspect. The port at Lothal served as a significant hub for trade between the Sumerian and Indus valley civilizations. From 600 BC onward The Babylonian and Egyptian empires are where the oldest forms of leisure travel may be found. In Babylon, there was an open-air museum displaying historical artifacts.

The Egyptians celebrated numerous religious holidays that drew a large crowd of worshippers and curious onlookers who flocked to the cities to view well-known artwork and structures. Like everywhere else, rulers traveled to expand their empires. Travel for religious reasons was done by both Brahmins and commoners. Numerous Brahmins and commoners flocked to Sarnath and Sravasti in large numbers to see the Buddha's inscrutable smile and be welcomed by him. The Greek civilization began in 500 BC. To the temples of the heaving gods went the Greek tourists.

The religious celebrations were also loved by the Greeks, who progressively turned them into occasions for recreation, especially athletic competition. For tourists going to the famous sights like the Parthenon, Athens has grown to be a key destination. To meet the demands of travelers, inns were built in significant cities and ports. The primary form of entertainment was provided by courtesans. Travel writing was also developed around this time.

The first travel writer ever was Herodotus. In the fourth century, travel manuals covering places like Troy, Sparta, and Athens also appeared. Advertisements in the form of placards pointing travelers to inns were also. Empire of Rome the conditions were favorable for passage because there were no external frontiers between England and Syria and because Roman patrols kept the seas free from piracy.

The expansion of travel was aided by first-rate highways and staging inns, the forerunners of today's motels. The Romans visited Egypt, Greece, Rhodes, Sicily, and Rhodes. Travel to the Holy Land also became quite common starting in 300 AD. The Romans created their travel guides, or itineraries, which included establishments with insignia to denote quality. The wealthy constructed second residences close to Rome, which they mainly used during the spring social. Around the Bay of Naples were the swankiest resorts. While Baiae, known for its rowdiness, intoxication, and all-night singing, attracted the lower class tourist, Cumae attracted the fashionable and the retired, and Naples attracted the retired and intellectuals. Tourism and travel would not share a similar position until the modern era.

Travel and tourism have become integral parts of our lives as it refreshes the mind and reduces stress. Every person on this world has his own likes and dislikes. The diversity of tourism ensures that there is something for everyone to do. Individuals think in different ways and human thoughts are diverse, as are human wishes and desires.

There are numerous reasons for people to travel. Some people travel for business, some for medical reasons in order to receive quality medical care, some travel in order to visit a holy location, and some travel in order to experience the natural diversity. Tourism can be divided into the following categories based on the primary reason for travel:

a. **Archeological Tourism**

It is a different type of cultural tourism that seeks to encourage interest in historical archaeology and the preservation of historic sites. Similar to ecological tourism (or ecotourism), it is encouraged to foster the growth of cultural associations, businesses, and cooperatives that specialize in providing this type of service.

All goods connected with public archaeological promotion, including excursions to ancient sites, might be categorized as archaeological tourism. Museums, interpretive facilities, historical reenactments, the rediscovery of indigenous goods, festivals, or theater. Even though archaeological tourism is still relatively new, numerous international organizations and governments have already started to think about potential alternatives for the establishment of sustainable economic and social systems. Egypt, Mexico, and Peru are only a few of the nations that have long participated in initiatives modeled after the key components of archaeological tourism for financial gain.

b. **Cultural Heritage Tourism**

Cultural heritage tourism (or just heritage tourism) is a branch of tourism oriented towards the cultural heritage of the location where tourism is occurring. Travelers who engage in cultural tourism learn about the traditions of a nation or religion, particularly the way of life and other factors that have impacted the people's way of life.

The growth of the Grand Tour from the 16th century onwards attests to the importance of culture as a vacation destination. In the 20th century, some have asserted, culture ceased to be the goal of tourism: tourism is now culture.

Cultural attractions play a significant role in tourism at all levels, from the global highlights of world culture to attractions that support local identities. Heritage tourism includes visits to historical or industrial sites, such as old canals, railroads, battlefields, etc. The main goal is to develop an appreciation of the past.

It also refers to advertising a place to Diaspora residents who have distant family ties to that area. The main influences behind a lot of modern heritage tourism are immigration and decolonization. Increased access to historic tourism has also been made possible by declining travel expenses. Travel for religious purposes or pilgrimages are another option.

The Vatican City, as well as other places like Lourdes or Fatima, attract a large number of Catholics from around the world. Numerous Jews have immigrated to Israel as well as traveled there. Many people have also visited Holocaust memorials and locations. Islam requires its adherents to perform the Hajj at Mecca, which sets it apart from traditional tourism even if the pilgrimage can still be a significant cultural experience.

Historical occurrences that have been dramatized for entertainment purposes also contribute to heritage tourism. For instance, a historical tour of a town or city may be employed, together with themes like Vikings or ghosts.

c. Pilgrimage Tourism

A pilgrimage is a protracted travel or search that has significant moral weight in religion and spirituality. It may involve traveling to a shrine or sacred site that is significant to the person's faith and beliefs. Pilgrimages are undertaken by followers of all major religions. A pilgrim is a person who undertakes such a journey.

The four places of pilgrimage offered by Buddhism include Kapilavastu, the birthplace of the Buddha. The places where he first preached in Banaras and Bodh Gaya, where he attained enlightenment, and where he attained parinirvana at Kusinagara. Pilgrimages from many different religions gather in the Holy Land like the Baha'i Faith, Christianity, Islam, and Judaism.

Before the advent of modern technology, pilgrims played a significant role in long-distance trade and brought prosperity to popular pilgrimage destinations, an economic phenomena that was unmatched until the development of the modern tourist industry. Assembling (and occasionally inventing) relics and penning hagiographies of local saints with moving narratives of miraculous healings were both done with the intention of inspiring pilgrims.

This spirit is preserved in Lourdes and other contemporary pilgrimage sites.' The Hajj, or trip to Mecca, is one of Islam's Five Pillars. All able-bodied Muslims who can afford to do so should attempt at least once in their lifetime. It is the most significant pilgrimage for Muslims. Ziyarat, a pilgrimage to locations connected to the prophet Muhammad, his companions, or other revered people in Islamic history like Shi'a imams or Sufi saints, is another common practice among Muslims.

Mosques, cemeteries, battlegrounds, mountains, and caves are some of the pilgrimage locations. Local pilgrimage customs, such as Ziyarat visits to nearby tombs, are also prevalent throughout Muslim nations. As visiting the graves during auspicious periods is a display of national and community identity, certain nations have particularly significant Ziyarat traditions surrounding the 14 hero grave sites.

d. Adventure Tourism

Adventure tourism is a type of tourism that involves exploration or travel to distant, exotic, and possibly hostile locations where travelers should "expect the unexpected." It involves engaging in physical activity while also learning about a new culture or landscape. As travelers look for alternative sorts of vacations, adventure tourism is quickly gaining popularity.

The global Adventure Travel Trade Association, which is based in the United States, defines "adventure travel" as any tourist activity that has two of the following three elements: a physical activity, a cultural exchange or encounter and participation in nature. Allowing participants to push outside of their comfort zones is one of the ways adventure tourism generates its thrill.

This could be as a result of suffering culture shock or by engaging in risky or labor-intensive behavior (real or perceived). Some of the adventure tourism activities involve day hiking, backpacking, zip lining, free fall, rafting, mountain biking, skiing and snowboarding.

e. **Agri-Tourism**

Agri-tourism, also known as agro-tourism, frequently takes place on farms. Travelers can get a taste of authentic local cuisine, experience rural life, and learn about various farming duties. This can involve the opportunity to assist with farming duties while you're there. In the wine-growing regions of Italy and Spain, agri-tourism is a common practice.

Agri-tourism is very common in America and encompasses any farm that is accessible to the public for at least a portion of the year. Visitors can select their own fruits and veggies, ride horses, sample honey, and learn about wine, visit farm stands and gift shops for locally grown food or handcrafted items, and much more.

Agri-tourism is quickly growing into a significant portion of the tourism business and will soon rank among its largest subsectors. People are more curious about how their food is made and want to interact with producers to learn more about how food is made. Children that visit farms frequently haven't seen a live duck, pig, or apple straight from the tree.

The term "entertainment farms" refers to the development of this type of expanded agri-tourism. With cuisine, mazes, open-pen animals, train rides, picnic areas, and pick-your-own fruit available, these farms cater to the pick-your-own crowd in addition to providing conventional farm items.

f. Atomic Tourism

An innovative form of tourism known as "atomic tourism" involves visitors visiting important sites in atomic history. Recent years have seen the emergence of atomic tourism, in which visitors visit locations like museums housing nuclear weapons, missile silos, and vehicles that once carried nuclear warheads to learn about the atomic age.

g. Bookstore Tourism

Bookstore tourism is a type of cultural tourism that promotes independent bookshops as destinations for tour groups. It began as a grassroots initiative to help locally owned and run booksellers, many of whom have had difficulty competing with big-chain stores and internet sellers.

h. Disaster Tourism

Disaster tourism is the act of traveling to a disaster area as a matter of curiosity. Visits to locations where there have been natural or man-made environmental disasters are referred to as disaster tourism, gloomy tourism, or grief tourism. People are drawn to this kind of tourism because of its social, academic, or cultural components.

If the conduct interferes with efforts to provide relief and healing, it may be bothersome. Some of the most famous disaster tourism spots in the world are Hurricane Katrina, Pompeii, Exxon Valdez Oil Spill, Naval Air Engineering Station Lakehurst, Nevada National Security Site, Johnstown Flood National Memorial and 9/11 Memorial.

In the wake of Hurricane Katrina, disaster tourism began to flourish in the Greater New Orleans Area. There are now escorted bus trips of areas of neighborhoods that suffered significant flooding due to storms. Some locals have considered these trips as unethical because the tour operators are making money off of their communities' and families' suffering.

i. Excursions

An excursion is a journey made by a group of people, typically for fun or learning. It frequently serves as an add-on to a lengthier trip or visit to a location, occasionally for other (usually professional) reasons.

For the purpose of luring this kind of business, public businesses offer discounted trip tickets. For the destination in question, these tickets are frequently limited to off-peak days or hours. The term "trip" refers to a brief excursion for learning purposes or to observe a natural phenomenon.

j. Garden Tourism

A specific sort of tourism known as "garden tourism" involves visiting or traveling to botanical gardens and other locations relevant to the history of horticulture. In familiar countries, garden visitors frequently travel independently, but they also choose to join organized garden tours in unfamiliar nations where they may have trouble speaking the language, getting around, or finding lodging close to the gardens.

Famous gardens which attract tourists around the world are Sissinghurst Castle Garden, Stourhead, Versailles, Giverny, Villandry, Rivau, Keukenhof, and Villa d'Este, Villa Lante, Alhambra, Longwood Gardens, Filoli, Taj Mahal and Ryōan-ji.

k. Medical Tourism

The phrase "medical tourism," also known as "medical travel," "health tourism," or "global healthcare," was first used by travel firms and the media to describe the fast expanding trend of people traveling across borders to receive medical treatment. Travelers who go abroad for medical care are said to be engaging in medical tourism.

These services mainly consist of elective procedures as well as complex specialist surgeries including joint replacement (knee/hip), heart surgery, dental surgery, and esthetic surgeries. With less formal legal or regulatory control to ensure quality and less formal recourse to reimbursement or restitution, as necessary, the provider and client use informal channels of communication, connections, and contracts.

The high expense of medical care, lengthy wait times for some procedures, simplicity and affordability of international travel, and advancements in technology and care standards in many nations are some of the factors that have contributed to the rising popularity of medical tourism.

l. Eco-Tourism

In its purest form, tourism is an industry that strives to have little negative impact on the environment and culture of the region it visits while also assisting the community in providing income and employment for locals.

Ecotourism entails visiting untouched natural places. It encourages environmental protection and enhancing locals' quality of life. Some of the best ecotourism destinations in the world are Alaska, The Amazon, Antarctica, Australian Blue Mountains, Borneo, Botswana, Costa Rica, Dominica, Galapagos Islands and Himalayas.

m. Geo-Tourism

According to National Geographic, geo-tourism preserves or improves a location's unique geographic characteristics, including its environment, cultural history, visual appeal, and overall wellbeing of its inhabitants.

Top geo-tourism destinations in the world are Tennessee River Valley, Sierra Nevada. Scenic, Wild Delaware River, Heart of the Continent, Crown of the Continent, Lakes to Locks Passage, Sedona Verde Valley, Four Corners and Bahamas.

n. Space Tourism

Traveling to space for leisure is known as space tourism. Despite sounding futuristic, something has already made history. A Russian Space Agency has already facilitated successful orbital space tourism. The different types of space tourism are orbital, suborbital and lunar.

International tourism is expected to continue to expand, with yearly growth rates varying from 3 to 6% depending on the destination. This industry, which is among the biggest and fastest-growing in the world, will continue to expand, putting significant strain on indigenous cultures and remaining areas with biological diversity, which is frequently utilized to sustain mass tourism.

Tourists who support sustainable tourism are aware of these risks and work to safeguard tourist locations as well as the tourism sector as a whole.

1.6. EXPERIENCE IS CAREER LADDER

There are management positions available in every area of the hospitality industry. A manager with multi-unit responsibilities for a restaurant, pub, hotel, club, or catering outlet is considered as being vitally essential because businesses in the hospitality sector operate multiple units.

There is also a need for a lot of multiunit or area managers when there are a lot of tiny units, such the restaurant and contract catering industries. Specialist managers, such as those in HR, marketing, or finance, are frequently hired at the head office level rather than within the divisions themselves to handle the large size and complexity.

In these types of managerial roles, there are numerous career options available, although they vary greatly by industry. Its size and complexity rank among the major considerations. Smaller units in restaurants, hotels, contract food outlets, or pubs typically have unit management that is active and focused on operations.

The general manager of larger enterprises, like hotels, is increasingly expected to concentrate on longer-term strategy, product, and market development, while his or her deputy is responsible for managing the day-to-day operations.

1.7. TOWARD A NEW PARADIGM

Institutions of higher learning are involved in educating students to be prepared to handle the challenges of innovation. Institutions now develop long-lasting educational programs that give students the knowledge and abilities needed for the corporate world of today and tomorrow. Young students require a smooth progression from being qualified to becoming competent.

Industry-inspired schools that have developed and adapted to higher education requirements are more prevalent in some countries than in others to empower them. It is also remarkable that certain nations do not appear in the top rankings, despite the fact that the economic importance of their hotel industry would undoubtedly warrant the presence of a robust hospitality education system in those nations.

Many schools have redesigned their practices or recontextualized them putting the focus on managerial experiences. The fact that students participate in operational activities is sometimes seen as undesirable in view of university standards and regulations, whilst a lack of preparation for practical work will generally jeopardize industry relations and graduates' options for career entrance.

One of the driving forces behind this approach that favors theoretical education above practical education is the advancement of the educational system and institutions. Since practical facilities are costly, budget constraints have forced schools to either reduce the appropriate instructional experiences or to outsource them, for example, by working with a for-profit company.

The competitiveness of "academicized" curricula has been impacted by these measures since prospective students evaluate the availability of practical facilities while making their study choices (Figure 1.2).

Figure 1.2. Global Hospitality education expo.

Source: Image by Zista Education

1.8. SCOPE AND FUTURE WILL NOT DISAPPEAR BUT TAKE DIFFERENT FORM

The incorporation of hospitality programs into academic settings, the discipline has matured to the point where schools and universities are now considered sources of knowledge, as opposed to when it was first developed in other places, such as business settings, governmental studies, think tanks, etc.

However, this maturity has also made these programs vocational foundations a restriction. Institutions have restricted the subject's growth to a discussion of the larger issues that are raised by and underlying the hospitality industry by tying curriculum development too closely to industry needs.

Economically, the travel and tourism sector alone provided an astounding \$8.9 trillion and created 319 million employments to the worldwide economy in 2018. In fact, the growth rate of this sector is actually higher than the combined rate of global GDP. This dictates volumes about the significant contribution to the global economy that the travel and tourism sector has been providing on an ongoing basis. A wider range of students seeking time management and organizing abilities are drawn to careers like hotel manager, chef, and accommodations manager. These management positions in the hospitality industry are both challenging and rewarding. Because there are so many options, the career growth in this field is extremely quick.

1.9. FUTURE OF HOSPITALITY EDUCATION

Digitalization is one of the most cutting-edge trends in hospitality education. The transformation of the educational system is just one example of how digitalization is connecting people, cities, countries, and disciplines of knowledge in new ways.

Since the world is increasingly becoming a place where everything happens online, educational institutions must adapt to embrace new learning styles and technology that can attract students. This is the fundamental change that the educational sector is undergoing.

The COVID-19 problem served as a catalyst for the digital transformation in the field of education, which had been anticipated as a trend influencing education but has since evolved into a necessity. In order to handle complicated situations, rethink business models, and support a sustainable future, future hospitality will need to rely on emotional intelligence.

We make sure to incorporate the finest of both the "real" and the "virtual" worlds into our educational techniques to provide them the best experience that is specifically tailored to their needs as well as the abilities needed to excel in their work and in life.

1.10. BENEFITS OF PERSONALIZED E-LEARNING MODELS FOR TOURISM AND HOSPITALITY EDUCATION

It is possible to learn anywhere and at any time due to the webification of education. As a result, the Internet provides a lot of flexibility to accommodate the unique working circumstances within the tourism and hospitality industries. Therefore, e-learning offers flexibility in terms of the time and place of instruction delivery as more students studying tourism and hospitality simultaneously seek out part-time jobs (Baum and Sigala, 2001).

E-learning models can be used by tutors to improve and assist students' learning while they are on placement, and they can also fulfill the needs of continual professional development, allowing people currently employed in the field to pursue lifelong learning while at work (Kasavana, 1999).

However, merely digitizing current instructor-oriented techniques does not result in increased benefits.

Contrarily, when collaboration and constructivism theories are used to utilize Internet capabilities for the creation of a student-centered, round-the-clock virtual classroom, the benefits to teaching and learning considerably increase. In this instance, learning is encouraged as a process that is fundamentally diverse, collaborative, and social rather than as a simple matter of memorization of pre-written information.

Indeed, the intellectual synergy of many minds, experiences and knowledge, benefit learners from: active and constructive learning, deep processing of information, improved individual achievement (Abrami and Bures, 1996); increased store of knowledge, improved communication and listening skills (Cho et al., 2002); the development of social attitudes and a collaborative spirit, motivation to learn, critical thinking, diversity of ideas and students' long term retention (Flynn, 1992).

Bernard et al. (2000) and Harasim (2000) also argued that online collaborative e-learning has the potential to overcome some of the problems associated with the distant student's out-of-campus situation, for example: isolation felt by the learner; high average rates of dropout; difficulties related to the formation of and entrance into the peer group of fellow students; low quality of learning attainment, such as the attainment of complex conceptual relationships and their associated skills; fear related to engagement in academic discourse; the time and place constraints for students dependent on employment; the trouble of building a shared understanding amongst students' multiple backgrounds and for students located in multiple cultures.

The acquisition of complex and higher level concepts and skills that have been claimed as weakness of traditional distance learning (Abrami and Bures, 1996; Harasim, 2000) the development of social, multicultural, and communication skills is seen as being of utmost importance for tourism and hospitality graduates.

According to Cho and Schmelzer (2000), online interaction with people from various social and cultural backgrounds allows learners to experience multicultural diversity and teamwork while also assisting them in becoming acclimated to the technological changes that are taking place in the tourism workplace.

This exposure can enhance their managerial abilities for the tourism and hospitality industry, which typically employs individuals from various cultural, racial, socioeconomic, and educational backgrounds, all of whom learn, think, and behave in different ways. Further, managers must competent to interact effectively via technology with their peers, clients, and employees who may work in various organizational cultures and settings due to the rising use of e-business, e-commerce, and strategic alliances between tourism organizations (Cho and Schmelzer, 2000).

Thus, e-learning can be a useful tool for educating students for the global tourist profession, which increasingly demands a skilled workforce capable of collaborating across disparities in location, time, and culture. E-learning offers excellent opportunities to learn how to utilize the Internet effectively and proficiently.

Graduates with the technical skills and expertise necessary to efficiently and creatively develop and adapt technological breakthroughs in corporate strategies and processes are highly desired, since technological developments drive constant change in the tourism and hospitality industries.

Additionally, since more and more travel and hospitality businesses are implementing online training and Intranets for knowledge management tasks, e-learning is thought to familiarize students with these settings and increase their competence in them by fostering their information literacy and knowledge management abilities.

The advantages of collaborative e-learning models, however, cannot be realized when the learning styles and preferences of an ordinary learner are taken into account. Further, many students lack the necessary skills to succeed in online learning environments, even while they should want and aim to become more independent of the teacher, self-supporting, and self-directing.

When this occurs, online forums may become a monolog of lecture-style content with few responses, conversations may evolve into a disorganized mountain of knowledge that is confusing and overwhelming for the participants, and a high volume of messages may result in information overload.

As a result, acknowledging the learning capacity gap among online students and offering solutions that take a holistic view of the individual are important first steps in assisting students in making the transition to online learning that is more successful, self-directed, and collaborative. Finding out what motivates students to want and intend to improve their performance is very crucial.

Identification of elements influencing individual learning processes is critical when taking into account the different profile and needs of learners attending tourism and hospitality education (such as high school graduates, persons with industry experience, and industry professionals). Personalized e-learning models emphasize the value of adapting instruction to individual variations in order to support learners in becoming more sophisticated, self-motivated, and self-directed learners.

This is because personal learning variances in how people learn have been identified. In today's knowledge-based society, information is not a scarce resource; rather, its application is. Every learner should be able to gain specific sorts of knowledge ability (combinations of skills and knowledge) using a personalized e-learning model in order to ensure his or her employability in the knowledge era.

Such types of combinations would be:

a. establishing linkages between theoretical and practical learning methods to allow for the integration of formal education into the workplace;
b. learning how to communicate with others in learning communities and use information and communication technologies as a resource to develop new knowledge; and
c. Developing a transformative rather than an informative relationship with the world.

In other words, knowing how to mediate and debate ideas or concepts that might serve as the foundation for new approaches to tackling the variety of economic, technological, and social changes instead of relying on established paradigms to interpret problematic circumstances

1.11. CONCLUSION

This chapter summarizes tourism and hospitality education. Chapter includes the introduction of tourism along with different types with tourism. In the mid of the chapter, topics like hospitality management career paths including early education, changing styles, combination styles are discussed.

Moving towards the end of the chapter, topics like experience as a career ladder, moving towards a new paradigm and scope and future will not disappear but take different form are summarized.

REFERENCES

1. Atef, T. M., (2018). A tourism and hospitality education management model: The case of the tourism department of sultan Qaboos university. Journal of Tourism and Hospitality Management, 6(1). Available at: https://doi.org/10.15640/jthm.v6n1a5 (accessed on 14 June 2023).
2. Cooper, C., et al., (2022). The Sage Handbook of Tourism Management. SAGE Publications Inc. Available at: https://us.sagepub.com/en-us/nam/the-sage-handbook-of-tourism-management/book263162 (accessed on 14 June 2023).
3. Dam, D., (2022). Scope and Future of Tourism Management. Toronto School of Management. Available at: https://www.torontosom.ca/blog/scope-and-future-of-tourism-management (accessed on 14 June 2023).
4. Hospitality Education and its Trends in 2021 – ET Hospitality World, (2021). ETHospitalityWorld.com. ETHospitalityWorld. Available at: https://hospitality.economictimes.indiatimes.com/news/speaking-heads/hospitality-education-and-its-trends-in-2021/80766369 (accessed on 14 June 2023).
5. Innovation in Hospitality Education and Career Opportunities, (2021). The Education Magazine. Available at: https://www.Theeducationmagazine.com/innovation-hospitality-education-career-opportunities/ (accessed on 14 June 2023).
6. The Changing Landscape of Hospitality Business Education: Past, Present and Future, (2022). ETHospitalityWorld.com. ET Hospitality World. Available at: https://hospitality.economictimes.indiatimes.com/news/speaking-heads/the-changing-landscape-of-hospitality-business-education-past-present-and-future/95348403 (accessed on 14 June 2023).
7. Tourism Management – Factors Affecting (n.d). Tutorials Point. Available at: https://www.tutorialspoint.com/tourism_management/tourism_management_factors_affecting.htm (accessed on 14 June 2023).
8. What is Hospitality Management? (n.d). hospitalityinsights.ehl.edu. ehl.edu. Available at: https://hospitalityinsights.ehl.edu/hospitality-management (accessed on 14 June 2023).

Chapter

2 TOURISM MANAGEMENT: APPLICATIONS OF THEORIES AND CONCEPTS OF TOURISM

Contents

2.1. Introduction 30
2.2. Concept and Functions of Management 31
2.3. The Concept of Tourism and its Definition 39
2.4. The Pillars of Tourism Theory 42
2.5. Tourism Theory Applications 46
2.6. General Systems Theory and Tourism 48
2.7. Conclusion 54
References 55

This chapter talks about the applications of theories and concepts of tourism in tourism management. At the beginning of this chapter, this chapter explains concepts and the functions of management. This chapter also addresses the concept of tourism and the definition of tourism. This chapter provides highlights on the pillars of tourism theory and on the tourism theory applications. At the end of the chapter, this chapter also explains virtual tourism and general system theory.

2.1. INTRODUCTION

Understanding management, how it works, and the responsibilities of managers is always helpful, regardless of whether your career goals are to manage, own, work for a tourism organization, or be a professional in the field of tourism.

Each person aspiring to work in any area of tourism must be aware of the fundamental management principles and practices in order to be successful. The promotion and development of responsible tourism requires a professional approach for managing various operations, and this knowledge is crucial for everyone aspiring to work in the industry.

To give you the necessary management skills and information, an effort is made in this course to familiarize you with various management-related challenges.

It aims to familiarize you with management principles and concepts as they relate to a variety of tourism-related operations, teach you the necessary skills, introduce you to the idea of motivation and the main theories of motivation, and help you understand different leadership styles by becoming familiar with some key theories of leadership.

This session also covers a brief examination of both individual and collective behavior. Since tourism is a social, economic, and cultural phenomenon that satisfies a fundamental human need, it has grown significantly in importance as a means of improving the conditions of human life, whether they be behavioral, social, or cultural, on a theoretical or practical basis.

The human being, in order to nourish himself philosophically, spiritually, culturally, and socially, and in order to aid him in purifying the senses, expanding perception, and growing knowledge, as well as represent him as a means of intellectual communication between peoples to exchange moral and social values and learn about cultures.

In addition to the ongoing expansion of world tourism and the potential for future space travel, customs and traditions are also important.

The purpose of this research is to develop a cognitive and scientific diagnosis in the direction of studying and examining tourism, which has a philosophical need to support its scientific theory based on the major intellectual pillars on the subject of tourism theory and the nature of the foundations for its formulation.

And presenting his issues as an effort to ground the intellectual discussion based on the complexity of its structural variables, its representation in reality, and the interactions and exchanges between the Center for Tourism and Complementary Activities.

The nature of the conceptual gap connected to the components of the emergence of tourism theory, the level of the conceptual agreement among scholars, and the nature of this gap, which still suffer from consequences that cause instability in the intellectual dimension.

The researcher's knowledge of various research attempts that focus on establishing tourism thought within a philosophical system based on building a theory of tourism, verifying the interconnected dimensions and explaining its relations within the mechanisms of building a theory of tourism, and the research comes as an attempt to put forward the theory of tourism is limited by these research attempts' differing philosophical structures and the absence of the scientific balance.

2.2. CONCEPT AND FUNCTIONS OF MANAGEMENT

In today's society, management has taken on significant importance. Its importance in the contemporary industrial world is growing daily. All organizations, whether commercial or nonprofit, civil or government administration, military or medical establishments, clinics, schools, or even domestic settings, require effective management.

It is important for us to have an effective and efficient management system in place so that we can use all of our material, physical, and human resources appropriately and easily reach the goals we have established.

All of the actions a manager does to achieve the goals of the business are considered to be part of management. Although many authors have defined management in a variety of ways, the theme of management remains the same.It involves creating and sustaining a setting so that predetermined goals can be achieved effectively. Planned, organized, staffed, leading, and controlling tasks are performed by managers. It emphasizes how to manage people well at work and to point them in the right direction so that they would be inspired to work towards achieving organizational goals.

It suggests making the best possible use of financial and physical resources in order to cut costs and increase profits. Henry Fayol defines management as "to manage is to forecast and plan, to organize, to command, to co-ordinate and control" (Figure 2.1). S. George argues that, "management consists of getting things done through others. A manager is one who accomplishes the objectives by directing the efforts of others."

Figure 2.1. Henry Fayol.

Source: Image by Store norske leksikon

According to George R. Terry "management is a distinct process consisting of planning, organizing, activating and controlling performed to determine and accomplish the objectives by the use of people and resources."

Management is a three tier activity. The top tier focuses on setting goals and policies, while the middle tier is concerned with putting those policies into practice with the help of the organization's lower tier.

The board of directors, its chairman, the managing director, general manager, and other important officers make up the top level of management.

This is the highest level of authority entrusted with running an enterprise's activities judiciously, successfully, and satisfactorily by careful planning, meticulous regulation, setting up efficient execution, and vigilant control.

It serves as the primary mobilizer of all human, material, and physical resources required for the undertaking's conception, upkeep, and growth. The duty of putting top management's outlined goals and policies into action falls under the purview of middle level management.

It often includes departmental heads and other executive officers who are affiliated to several departments. Lower level employees are supervised by middle level management, who also utilizes them to complete job.

Executives at this level of management are responsible for supervising and directing operational staff. It includes managers, foremen, financial and accounting officers, sales representatives, etc.

2.2.1. Objectives of Management

The objectives of management can be classified as follows: Organizational: The basic goal of management is to effectively utilize both material and human resources in order to maximize benefit to the business.

Every management group strives to achieve these objectives. These objectives can be profit, survival and growth.

- Survival: Every organization has survival and perpetuity as its primary goals. Through appropriate profits, management must ensure survival.
- Profit: Every business's primary goal is to maximize profit. Management must therefore make it possible for the company to maximize earnings.
- Growth: In order for their business to grow and expand, organizations must constantly expect to make a profit that is sufficient and to last for a long time. Management must assist in addressing all dangers, uncovered elements, and upcoming obstacles in order to guarantee growth and prosperity.

Social: Management must adopt a social perspective and address social issues through welfare programs. An organization's social goals include adopting environmentally friendly production techniques, offering customers high-quality goods at competitive costs, giving people access to employment possibilities, and providing employees with basic amenities like access to healthcare, childcare, and schools.

Personal: People join groups in order to accomplish their own goals and meet their basic requirements. They work to meet their basic requirements, such as salary and other benefits like bonuses, as well as higher-order needs, such as personal development, advancement, and future prospects.

Management must combine personal aspirations with organizational goals in order to provide harmony to organizations.

2.2.2. Importance of Management

The importance of management can be highlighted in the following ways: Management provides appropriate guidance for achieving shared personal and organizational objectives. All resources, including financial, material, and human resources, are useless without management.

Management maximizes the organization's goals by utilizing all other resources. An organization can adapt to a changing environment with the help of management.

A number of factors have a significant impact on business, including changes in technology, consumer preferences, fashion, and product demand. The management must adapt to these upcoming problems.

Management cultivates human values inside the company. The attainment of organizational goals is facilitated by men, not by materials, equipment, or resources. Management aids in boosting productivity to ensure an organization's success.

It effectively uses the time and resources that are at hand, cuts down on waste of any kind to cut expenses, and maximizes profits. An organization has several responsibilities to the various social groups that make up its membership.

The process of achieving their goals aims to advance both the company and society, offer better products at competitive rates, create job opportunities, boost national wealth, and raise worker wages.

2.2.3. Functions of Management

In general, management is viewed as a series of process, and the process refers to the numerous tasks that a manager performs in order. It is a process of using authority and taking on responsibility in order to plan, direct, control, carry out, and realize the outcomes of company activities.

All functions can't be performed by a single individual. According to Allen, it is humanly impossible for one person to adequately handle the numerous and intricate requirements of managing and running the organization. They must be entrusted to various people throughout the entire business.

The main functions of management are as follows:

a. **Planning**

The most important role in management is planning. Planning is known as the management function at the core, and it serves as the foundation for all other management activities. The process of defining goals, creating a plan to achieve those goals, and creating plans to integrate and coordinate operations are all parts of the planning function.

Planning ensures efficient use of both human and material resources to generate profits. It entails planning ahead to decide what to do, when to do it, how to do it, where to do it, when to do it, and by whom.

The nature and scope of planning varies as per the level of manager. For instance, an escort may plan for the group's entire tour or just the day's excursion, but the tour operator will plan for how the tour will be packaged, promoted, catered to, etc.

It is crucial that employees in a service firm, such as tourism, organize their actions while keeping in mind the organizational objectives, in addition to top and middle managers. It takes a lot of planning to implement any tourism-related service, development, or activity in a specific area, region, or level.

b. **Organizing**

Determining what tasks need to be completed, who will complete them, how the tasks should be categorized, who will report to whom, and at what level decisions are made are all part of the organizing process. It relates to how work is distributed among team members so that the organization's objectives can be successfully met.

Organizing is the systematic grouping of resources and activities to facilitate the achievement of particular organizational goals. While it is feasible to accomplish goals without formally organizing, there is a high possibility that this will result in significant resource and time waste.

Organizing ensures that goals are accomplished as quickly as possible, in a systematic manner, and with the best use of the available resources. It involves assembling of raw materials, equipment, mobilizing money, choosing staff, and figuring out the roles and responsibilities of human resources in various departments at all levels of the organizational hierarchy.

The elements of organizing include grouping of activities, distributing responsibilities, establishing fixed deadlines, delegating authority, and building relationships between various sources of authority. The resources at a firm's disposal include its employees, equipment, buildings, factories, money, and credit that are available for usage.

All these resources are limited. To ensure that there is no confusion, conflict, duplication, or wastage in attaining the organization's goals, it is the responsibility of the management to arrange all of these resources.

The management will be expected to assign each employee their duty, designation, and position within the company, as well as their authority to use the resources that have been given to them and the higher authority to whom they are required to report their progress on a regular basis.

c. Staffing

The most valuable resource and most valuable asset of a firm is its human resources. An organization's staff is just as crucial as its physical structure. Staffing includes hiring, choosing, training, placing, and promoting employees.

Planning the workforce, analyzing jobs, and performing other staff duties are necessary. It is an ongoing process due to personnel attrition, retirements, and new requests due to expansion, etc. A company's motto should be "right man for the right job at the right time and doing the right things" if it wants to succeed and meet its objectives.

The tasks involved in identifying a person's needs in relation to the work to be done are included in the staffing function. It is necessary to follow the correct choosing process. Sensible selection should be made.

After hiring the appropriate and necessary staff, it is important to effectively train them and help them adapt to the changing environment of the organization. Therefore, care must be made when selecting the best human resource.

Depending on the size of the organization and the importance of the tasks, either all managers in the company or a separate Department called the Personnel Department fulfill this duty.

d. Directing

Directing means telling others what to do and ensuring that they carry it out as effectively as possible. It aims to direct and motivate employees to carry out the task in the best manner feasible. The execution of planning is what direction is all about.

The organization is given new vitality and organized activity is started. Getting things done through and with people is a key component of direction as a management strategy. To function properly, they need constant support and encouragement.

After receiving the best orientation and training, it is the responsibility of superiors to mentor, inspire, and lead the workforce toward improved performance. If instructions are flawed and inadequate, even the best planning and organizing may fall short of expectations.

e. Coordinating

The art of coordination involves bringing together the activities of individuals and groups to accomplish shared objectives. All other managerial functions are said to be obligated by it. It is a procedure for incorporating the teamwork of individuals into an organization.

The goal of every managerial process is considered to be coordination. It is the seamless balancing of all the manufacturing elements. Employees of an organization cannot successfully accomplish the goals of their interest if they lack a sense of teamwork, just as players in a hockey or cricket team cannot defeat the opposing side without adequate coordination.

According to Henry Fayol, "to coordinate is to synchronize all of a concern's activities in order to facilitate its operation and success." Planning and execution, no matter how precise and skilled, cannot be successful unless management creates a harmonious blending of activities through organizational structure and administrative action.

The administrative set up needs to be in place so that each stage's actions are carried out in the appropriate order and the desired outcome is achieved in the end. The same goes for activities related to production, purchases, sales, finance, etc., which are combined to maintain continuity.

f. Motivation

Motivation is a strategy used by management to encourage employees to perform harder and more effectively. It is a motivating factor that encourages all of an organization's staff to give the organization their all in order to accomplish the predetermined goals of the business.

It is a psychological tactic for carrying out the plans and policies with the help of others. The management process will not be finished by hiring a worker and giving him a job description. There is a saying that goes, "You can bring a horse to water, but you can't make him drink."

With regards to management's goals and directions, motivation is a positive force. Similar to this, a manager must remember that while he may purchase a man's time and physical presence at work, he cannot purchase a man's enthusiasm, loyalty, or initiative.

Management must inspire employees to want to use their talent for better performance.

g. Controlling

Controlling is referred to be a yardstick for gauging an organization's performance. It is a procedure for reviewing and rating the work produced by subordinates in various organizational departments.

To make sure that things go according to plan, it involves monitoring and correcting the actions of subordinates. It evaluates performance in relation to objectives and plans, identifies negative deviations, and, by putting corrective measures into motion, helps to guarantee that plans are carried out.

Controlling enables us to make sure that all of an organization's actions are carried out in the manner that was first intended. Data of actual results are gathered during the control process and contrasted with those anticipated if the results are as expected.

Controlling requires management to confirm that performance of the results adheres to the plans, policies, and standards that were approved. In order to redirect the efforts of the subordinates back to the desired direction, it is primarily intended to identify flaws in work performance or deviations from the planned course.

The best managerial control addresses plan deviations before they happen since controlling is a process that looks backwards.

2.3. THE CONCEPT OF TOURISM A ND ITS DEFINITION

Man has known tourism since the beginning of creation as he lived in constant movement and search for stability and safety while wandering the earth and before knowledge advanced, social and cultural ties were established, the Sumerians and Babylonians minted money, and their discovery of writing helped to significantly increase trade and freedom of movement (Figure 2.2).

Figure 2.2. Giza Pyramids.

Source: Image by Flickr

The ancient Egyptians also managed to develop the first tourist fleet, which was intended to travel to Africa in order to spread peace.

Whereas the movement of travel, because of its cultural and religious history, represents the fundamental structure of an ancient civilization, tourism took on a philosophical concept because it was seen as a triad of the mind, body, and soul and it evolved into a professional concept with the idea of a numerical balance between knowledge, wisdom, and craft.

The khan was the foundation of the tourist sector in ancient civilizations like the Greek, for instance, as it offers services to the visitor and his journey from lodging, food, and drink during his stay, as well as meeting the demand of the travel route.

The need for love of aspiration and discovery came as motives present in the human being in his movements and his trips between cities, along with it, in addition to the variety of purposes that made the Khan a place of residence in every part of the world.

According to the travelogs of Ibn Battuta, Magellan, and Marco Polo, which It can be counted from tourism activities, even if it does not involve under the current concept and meaning of tourism whose goals and directions differed, as the word tourism in its current meaning in foreign languages was not known at the time, as it is a modern word that was commonly used in the eighteenth century, the term (tour) was used to express (Figure 2.3).

Figure 2.3. Ibn Battuta Portrait.

Source: Image by World History Encyclopedia

The phrase was created by referring to the journey that every respectable young Englishman does to Europe in order to finish his studies by calling the young Englishmen who were taking it bass M (tourists) (Al-Dahir & Elyas, 2001).

As a result, many other languages adopted the word "tourism" in its meaning from the French language. The phrase was then used in France to describe anyone who was traveling for personal enjoyment. The only voyage without money that causes difficulties is one taken for leisure, rest, or medical treatment (Mahmoud, 1975).

The term has gained widespread usage as a result of a rise in exploration excursions around the world for the sake of acquiring economic and natural resources, as well as for education and interaction with nature, culture, industry, and people. Through their scientific experiments, they communicated their ideas about tourism.

As a result, the idea of tourism has continuously evolved to reflect its significance as a human phenomenon, an economic resource, and a fundamental social need that takes precedence in light of the complexity of life conditions.

It also reflects all relationships resulting from the actions of people who travel to locations outside of their home nations and live there for a period of time that does not exceed one year consecutively, whether for leisure, business, or other reasons. (2009 McCabe, 2009).

The early stages of the development of modern tourism started with the efforts of Thomas Cook in 1841, which saw the first announced journey with a train carrying 570 passengers from Leicester to Lorborough in England. Tourism activities then expanded during the Second World War, marking the beginning of the era of mass tourism.

Since they have been the primary component in the marketing and sale of a variety of tourism products for several decades, travel agencies have grown in importance as components of the tourism distribution chain. Rather than providing information about these products and tourist destinations, they have instead sold travel packages, transportation tickets, attractions, housing, and travel insurance (Lohmann & Netto, 2008) and by doing so, they have gained increasing importance.

One of the characteristics of the modern era, tourism justifies the development of international relations, the exchange of common interests, the creation of international institutions and organizations concerned with travel and tourism, as well as its active role in constructing the economies of countries and their development plans (Al-Najjar, 2001), as well as current account terms (Khan & Khan, 2006), and links to popular culture as a historical phenomenon. (McCabe, 2014) Just like when purchasing tickets for travel and lodging (Figure 2.4).

Figure 2.4. Marco Polo traveling.

Source: Image by collection-GetArchive

2.4. THE PILLARS OF TOURISM THEORY

Science has had a significant impact on many facets of life and was reflected in human thought (Al-Asmari, 2012), as it included many theories that served modern science and provided justifications and clarifications of the phenomena and events that interested them. These theories vary depending on their purpose, including normative theories, explanatory analytical theories, and descriptive theories. Other theories are categorized under.

In addition, research into the tourism phenomena is still relatively new compared to other areas of study including sociology.

At the start of the nineteenth century, pioneers in Europe tackled the topic of the phenomena of tourism, particularly German studies (Nash, 2007), which grew in popularity in an increasing number of English studies after the end of the Second World War, as these studies were characterized by desolation and their lack of scientific methodology.

The theory is a set of interrelated structures and assumptions that clarify the relationships between a numbers of variables and aim to explain and predict a phenomenon. It also represents standard solutions and basic specifications to increase the efficiency of control, simplicity, quality and cooperation (Jakobs, 2006).The significance of the philosophical framework upon which the theory of tourism is built, with a view to identifying new entrances and visions inherent in the phenomenon of tourism, with a view to accessing the entrances to the topic of tourism theory, through the requirement for a deductive intellectual framework that addresses the phenomenon, and is governed by the consistency and harmony between its components and its relationship of ties, adopted by the world.

In order to understand the causes of the tourist phenomenon and how to control, influence, increase, or decrease it, the researcher must respond to a number of propositions and hypotheses that make up the subject of the phenomenon (Obaidat et al., 1984). The tourist theory may also be related to practice and to its scientific and practical application.

It is connected to a synthetic, descriptive, intellectual production focused on the application and effect in reality in order for it to acquire a new meaning, and as a result, it has numerous dimensions in its folds, including the following:

- Realistic dimension: It is concerned with the subject matter in which tourism theory handles its description and interpretation.
- The technical dimension is related to the total procedures and processes that surround its production efficiently and effectively.
- The value dimension relates to its sincerity and cohesion.
- The cognitive dimension is the primary means by which the experimental scientist uses knowledge of the laws governing phenomena.

With the expectation that there will be one billion tourists worldwide, the tourism sector has established itself as one of the fastest-growing, broadest, and most prevalent industries in today's world. This has inspired authors and researchers to connect their scientific endeavors with many of their philosophical and intellectual premises.

Tourism studies have the potential to be a hub for the development of theories across all human and scientific disciplines, but they have historically suffered from a severe lack of theoretical issues due to their subject matter. However, as a result of the discussions of researchers and academics from other fields, new ideas about tourism, such as eco-tourism, cultural tourism, mass tourism, etc., have emerged.

As it focused on dealing with the tourist movement in a variety of ways, such as the focused behavior of tourists, the reasons for travel, and the tourist destinations, without considering a philosophical elaboration into a logical explanation that deepens the concept of tourism and proves its theory, and the blog theory, which was primarily focused on the psychological factor of tourists, while Erik Cohen's theory was related to the behavior and context of tourists, and other scholars have explored topics linking different disciplines such as recreation, residence, or uniforms.

There are a number of historical monuments and artifacts, as well as alluring natural resources, and this connection between the specializations was not a scientific victory for the tourism perspective, but rather a pure field to prove other theories, such as in sociology, psychology, management, etc., and so these directions did not give particularity to the tourism perspective.

Rather, it was combined to meet the scientific objectives of the other specialties because tourism is a complex variety of systems in social and economic structures, natural environments, royal patterns, and relationships between the generation of travel and polarized countries (Burns: 2005), regardless of distortion of travel theory or its continued vitality.

And postmodernism serves as a prime example of the tendency of ideas, models, and notions about transferring a person from one place to another, as well as the commitment of time to such changes.

The origin, the particular circumstances that give rise to the concept, and the distance traveled can all be considered fundamental aspects that affect the nature and movement of ideas across the intellectual landscape. (Brown & Turley, 1997)

There are many factors that contributed to the delay in the development of the theory for tourism, including the fact that the majority of studies and researches in a variety of fields, including sociology, psychology, anthropology, city planning, administration, and economics, etc., had their roots restricted to the human experience from a western perspective and to contemporary trips using transportation, and travel companies, without first going through a methodology inherent in the tools of sound scientific research.

From an Eastern perspective, tourism was only possible through the literature of prehistoric voyages, which had different meanings and details that could only intersect in specific directions.

Scientists and researchers have come to varied conclusions on the various meanings and pillars of the theory depending on the discipline in which it is generally used. Providing the natural phenomena occurs and there is no conflict in its occurrence, the theory can serve as a form of explanation for how it happens.

Regarding tourist theory, it explains how these phenomena arise. They are generally subject to right and wrong, but the explanation and interpretation of the theory, which is dependent on a large number of experimental findings that can broaden and enrich thought from what things, have been tried and judged through the knowledge that the mind acquires, is what gives it its logical coherence. Its goal is to arrive at a law that explains the incidents of tourism phenomenon.

The goal of this is to develop tourist theory from existing hypotheses or creating ideas, which is important to support the theory and give it more and more importance.

The importance of tourism theory lies in the functions it performs in the field of human knowledge, which are as follows:

1. Tourism theory works to collect tourist facts, concepts and principles and arrange them in building a coordinated organization, which makes tourism activities meaningful and valuable.
2. Tourism theory provides an explanation and interpretation of a number of relationships, ties and events accompanying it.
3. Help in predicting many tourism activities and anticipating their occurrence or not in light of certain data and indicators.
4. Tourism theory directs scientific thought, scientific research procedures and processes, and mental reasoning to build it.

The value of scientific theory, on the other hand, might not be tightened by the strength of empirical evidence to prove the validity and strength of certainty, but rather lie in its flexibility and exposure to all possibilities, the possibilities of adjusting and predicting its results, the ability to develop and open to legitimate scientific facts, its mistakes, and its insufficient depth to study and understand phenomena. There are three important issues in the interpretation of tourism theory that following:

1. *Theory Unit:* It is predicated on the existence of a strategy plan, plans to address a problem pertaining to tourists, or a variety of current issues with regards to engaging in travel and tourism-related activities.
2. *The Fertility of the Theory:* the potential of tourism theory to broaden perspectives and raise a number of issues in fresh and original study disciplines, which helps the globe gain a fresh perspective on understanding and interpreting the phenomena of tourism.

3. *Theory Assumptions:* The capacity of the theory of tourism to form hypotheses independent of the issue supports the logic of the theory in explaining the phenomenon of its occurrence and growth with the passage of time and events.

2.5. TOURISM THEORY APPLICATIONS

There is no reason to doubt that the differences in the types of tourism necessitate matching various symbolic and mental images when conceiving. The types of tourism activities and their applications vary according to many indicators of environmental giving to the natural attractions of environmental resources, or urban construction, or the cultural, historical, and civilizational component, etc.

Tourists, in which each tourist seeks the appropriate type of tourism to achieve on the basis of the effect that he wishes to acquire in his psyche and apparent behavior, and the resulting moral values, or an understanding of intellectual philosophy, or economic construction, the goal of diversifying tourism activities and events is due to the differences in those inclinations and desires of tourists according to what they desire and aspire to achieve, and on the basis of this, applications for tourism can emerge in the construction of the tourism industry from residential, recreational, therapeutic, religious, scientific, and other projects, and the species is not required to overlap with any of them; this depends on the visitor's preferences and his decision to travel.

This leads us to the conclusion that there are several categories of tourism that are susceptible to growth because they are compatible with each other in terms of tourism-related activities, goals, ownership, and other factors.

We will address the classification of types of tourism as follows.

2.5.1. Tourism with Goals

Every type of tourism product attracts a specific segment of tourists, and the design of its services will be in accordance with what they need and wish to achieve. These activities and events include recreation, cultural, religious, economic, industrial, political, therapeutic, sports, and social activities.

The tourist type of most tourist groups is thus characterized by the fact that they do not play a significant individual role in its development but rather are a response to a particular need or a collective aim in itself.

2.5.2. Tourism Within the Scope of Travel and Nationality

The type of tourism in the range of travel and nationality is outside the political borders of the country, and in a legal manner that necessitates the presence of fundamental approvals from the visited country such as entry visas, or reciprocal international relations, with regard to holders of dual citizenship, or citizens of other countries, and representatives of companies. This type is also subject to the nature of political relations.

Domestic tourism is distinguished by drawing large numbers of families, and it provides a lot of money spent in this field, and it is linked in a specific season like summer or school holidays. Domestic tourism does not require complicated or many procedures, but it do attract large numbers of families and is characterized by attracting economic interests, intra-commerce, or religious seasons as in (Hajj) and (Umrah).

Colleges and tourism-related activities are concentrated in inner-city and seaside tourist zones, where they help introduce visitors to their country and its numerous civilized accomplishments and whose knowledge is a component of national culture and civilized behavior (Figure 2.5).

Figure 2.5. Hajj.

Source: Image by Wikimedia Commons

2.5.3. Tourism for the Type of Accommodation and Transportation

This type of tourism is distinguished by its reliance on the nature and style of accommodations, the use of transportation for tourist movement, and the timing of the tourist season. As the numbers of tourist's increases significantly each year, the majority of tourism activities take place during the summer months in most tourist destinations, and the tourist movement requires accommodations in a wide range of settings, this type of tourism is characterized by these dependencies.

The tourism industry was encouraged to identify different types of accommodations in accordance with the recognized tourist classification levels because the number of beds available is insufficient to fulfill the real demand.

2.5.4. Virtual Tourism

Virtual tourism is now a reality in many nations across the world and is utilized as an alternative to flying travelers from one nation to another in order to experience the natural resources, buildings, artifacts from history, and other attractions that country's tourism industry has to offer.

The concept of virtual tourism is around giving visitors a virtual tour of popular tourist destinations via a triple website. Dimensions model shrines and museums before applying the journey via a website. If travel agencies develop specialized software and applications and then promote them internationally, the benefits of virtual travel are primarily focused on generating financial returns.

Due to the formal barriers that nations place in the way of traveling across their borders in places like airports, ports, and land ports, virtual tourism has emerged as a substitute for travelers who wish to travel to other countries in order to experience their real-world effects and historical attractions.

Through a three-dimensional website that mimics museums, shrines, palaces, virtual cities, and public parks, virtual tourism makes it easier to travel about virtually in the many tourist destinations. Science may learn from its mistakes and constantly revise its theories and interpretations of the events and their environment.

2.6. GENERAL SYSTEMS THEORY AND TOURISM

It is impossible to pinpoint one person as the originator of general systems theory, however there is agreement in academic circles that one of the pioneers of this theoretical subject was the biologist Ludwig von Bertalanffy (1901–1972).

The following authors, each in their own specialized field, contributed to the development and dissemination of generic systems theory: Béla H.

Bánáthy, Talcott Parsons, Gregory Bateson, Heinz von Forester, Niklas Luhmann, Humberto Maturana, Francisco Varela, Howard T. Odum, Eugene Odum, Edgar Morin, and Fritjof Capra.

A "system," according to Bertalanffy (1967), is a collection of components that are working together. No matter the nature of the constituent parts or the relationships or forces between them, there are universal principles that apply to systems.

According to Beni (2001), a system is "a collection of pieces that interact to achieve a certain objective, in accordance with a plan or principle or a set of processes, doctrines, ideas or principles, logically ordered and sufficiently coherent to describe."

The main concept in this theory is "system," which can be interpreted as "any structured totality or entire" (Abbagnano, 1999). Despite the fact that systems theory is holistic, several scholars have emphasized that there are differences between the concepts of holism and systemize.

In his book 'Holism and Evolution' from 1926, Jan Christiaan Smuts developed the philosophical concept of holism.

Leslie (2000) asserts that tourism benefits from a holistic perspective because it encompasses all of the industry's essential elements, including travel, lodging, and activities at the destination. This is true whether or not the industry is viewed as a business sector or a field of study at the academic level.

According to Leiper (1995), Bertalanffy came to the realization that in order to fully comprehend living things, he needed to move beyond biology and incorporate information from other sciences. Bertalanffy proposed a theory containing categories of thought that could be utilized in various sciences after realizing that the principles of physics were distinct from those used in biology and society.

He proposed the development of a thermodynamics method that would address open systems, which are not in equilibrium, based on classical thermodynamics, which deals with systems in equilibrium. Systems theory and systems thinking are separated by Leiper (2000).

The former, which was utilized by Aristotle, Ptolemy, Copernicus, Galileo, Ibn Khaldun, and Vico, is a way of understanding objects (the concrete system) or ideas (the abstract system). Contrarily, systems theory is far more recent, having been developed in the 1930s and still evolving today (Figure 2.6).

Figure 2.6. Aristotle.

Source: Image by Flickr

The concepts of Bertalanffy have appeared in a number of books and articles. General System Theory as Integrating Factor in Contemporary Science and in Philosophy (Bertalanffy, 1968b) is one that has been distributed in the USA, although his most well-known work is General System Theory, which was released in the USA in 1968.

This book proposes an interdisciplinary generic systems theory that may be applied to open systems of any kind. Every living thing, according to Bertanlanffy, is fundamentally an open system since it maintains a constant influx and outflow of substances (Bertalanffy, 1968a).

The core principles of Bertalanffy apply to many other kinds of systems, including the human body, a nation's economy, a municipality's political structure, and travel in any location. In order to better comprehend and research each of these systems, general systems theory allows for either an analysis of the system as a whole, or an analysis of the system in its component pieces.

A system must have all of the following components in order to be considered complete: an environment (the system's physical location), units (its components), relationships (the relationships between the units), attributes (the quality of the components and the system itself), input (what goes into the system), output (what comes out of the system), feedback (control of the system to keep it functioning properly), and a model (a system design to facilitate its understanding).

According to Leiper (1995), the main advantage of generic systems theory is that it makes otherwise complex things easier to understand.

Systems theory has been invalidated, according to Phillips (1969), a sharp critic of it. He identified five of its primary flaws:

1. The failure of systems theorists to appreciate the history of their theory.
2. The failure to specify precisely what is meant by a 'system.'
3. The vagueness about what it means to be included in systems theory.
4. The weakness of the charges brought against the analytic or mechanistic method.
5. The failure of general systems theory as a scientific theory.

According to the literature, Raymundo Cuervo conducted one of the first analyzes of tourism utilizing Bertalanffy's general systems theory (1967).

In the book El Turismo Como Medio de Comunicación Humana, edited by the Mexican government's Department of Tourism to commemorate the International Year of Tourism (1967), Cuervo published the chapter 'El turismo como medio de comunicación,' in which he describes tourism as a 'well-defined set of relationships, services and facilities that are generated due to certain human movements' (Cuervo, 1967). For Cuervo (1967), tourism is a large set composed of the following subsets:

1. Transport by air, car, rail, sea, river, etc.
2. Lodging establishments, hotels, inns and hostels.
3. Travel agencies.
4. Tour guides.
5. Restaurants, cafes and other establishments from which a fluctuating population obtains food and beverage services.
6. Commercial establishments dedicated to selling souvenirs, travel items and other goods usually consumed by travelers.

7. Manufacturers of souvenirs and other goods usually consumed by travelers.
8. Artisans dedicated to producing typical handicraft objects.
9. Entertainment centers whose clientele is largely part of the fluctuating population.

Each of these subsets can be divided into other subsets, as in the example below (Cuervo, 1967):

1 = (A1, A2, A3, … An)

Where:

1 = transport

A1 = domestic air transport

A2 = international air transport

A3 = domestic bus transport

An = other types of transport

Cuervo's study from 1967 is predicated on the idea that tourism is a set whose purpose is communication.

To support his position, he asserts that although tourism is a kind of communication capable of disseminating information that is both helpful and constructive for advancing world peace, it can also be detrimental and negatively impact the harmony of interpersonal relationships.

From this perspective, the tourism industry should constantly operate as a facilitator of constructive conversation (Cuervo, 1967).

The most well-known tourism system on a global scale was created by Neil Leiper, who also offered a model with five components, including three geographical components:

1. the traveler-generating region (the origin or environment where the travelers usually live);
2. a transit region that connects the origin to the destinations; and
3. The tourism destination region. The other two elements are:
4. the tourist; and
5. The tourism and travel industry (tourist information centers, lodging establishments, etc.).

According to Leiper (1990), the interaction of these five elements is influenced by external environmental factors, which in turn have an effect on the system's interactions with various environments (such as the human, sociocultural, economic, technological, physical, political, and legal environments) and how travelers move through the transit region.

For instance, the tourist-producing area is home to both the travel agencies that arrange the journey for the visitors as well as the marketing and promotion firms that wish to affect the level of tourism. The generating regions and the destination regions can communicate through the routes of transport, distribution, and movement.

The accommodation industry and tourism attractions are also found in the destination region. The schema makes it obvious that the tourist experience starts in the generating zone, travels via the transit region, and ends in the destination region.

In Leiper's system, each component interacts with and is affected by the system that helps to create the tourism product.

The key benefits of Leiper's model, in the words of Lohmann (2004), are its simplicity and wide applicability, giving a handy tool to "visualize" the tourism system. Its benefits also include the following (Cooper, 1998):

- The ability to incorporate an interdisciplinary viewpoint because it is not just dependent on one discipline (or environment), while at the same time offering a structure that identifies the specific contributions of each environment.
- The possibility of using the model at any scale or level of generalization, from a local examination to the industry as a whole in the global sphere.
- The model's adaptability allows for both the examination of the five fundamental components of tourism—which were previously mentioned—as well as the introduction of various types of tourism.

This model also demonstrates how the interrelationship between tourism's various components, a key concept in the field, works. A holistic strategy for promoting tourism was also put forth by Sergio Molina in 1997, and his concept has gained widespread recognition in Latin America.

According to Molina (1997), a systemic view of tourism is made up of a number of components or subsystems that work together to achieve a single objective. According to Molina (1997), the following components or subsystems make up the tourism system:

- Superstructure: public and private sector organizations; laws, regulations, plans, and programs.
- Demand: tourists living in the country and abroad.
- Infrastructure: airports, roads, water supply networks, sewage, telephones, etc.
- Attractions: natural and cultural.
- Equipment and facilities: hotels, motels, campgrounds, trailer parks, restaurants, cafes, travel agencies, pools and tennis courts, among others.
- The hosting community: local residents directly and indirectly linked to tourism.

2.7. CONCLUSION

In this chapter on tourism management, it has been explored how to use theories and concepts of tourism with management functions, including goals, significance, and functions. In addition to general systems theory, this chapter added information on tourism with goals, tourism with scope of travel and nationality, tourism for the type of accommodations and transportation, and virtual tourism. Tourism is important in that it takes a holistic approach that takes into account all of the industry's key elements, including traveler movement, transportation, lodging, and activities at the destination.

REFERENCES

1. Akkaya, E., (2020). *Tourism Theory Concepts, Models and Systems.* Academia.edu. CAB International. Available at: https://www.academia.edu/42063458/Tourism_Theory_Concepts_Models_and_Systems (accessed on 14 June 2023).
2. Aljubori, S., (2022). *(PDF) Tourism Theory: Its Application and Criticism – Theoretical Study.* https://www.researchgate.net. Available at: https://www.researchgate.net/publication/358045953_Tourism_theory_its_application_and_criticism_-Theoretical_study-_salm_hmyd_salm_amd_jamt_bghdad (accessed on 14 June 2023).
3. Lohmann, G., & Panosso, A. N., (2017). *Tourism Theory Concepts, Models and Systems.* https://dspace.uef.edu.vn. CAB International. Available at: https://dspace.uef.edu.vn/bitstream/123456789/29243/1/Tourism%20theory%20_%20concepts%2C%20models%20and%20systems%20-2017-910.01r.pdf (accessed on 14 June 2023).
4. McKercher, B., & Prideaux, B., (2020). *Tourism Theories, Concepts and Models. goodfellowpublishers.com.* Goodfellow Publishers Ltd. Available at: https://www.goodfellowpublishers.com/free_files/Cover,%20contents%20and%20copyright-faf1220edcb5552499e708986e3bebf4.pdf (accessed on 14 June 2023).
5. *Module – 5 Tourism Management of Tourism Business* (n.d). https://nios.ac.in/. NIOS. Available at: https://www.nios.ac.in/media/documents/tourism_337_courseE/337_Tourism_Eng/337_Tourism_Eng_L14.pdf (accessed on 14 June 2023).
6. Pender, L., & Sharpley, R., (2005). *Management of Tourism.* eiilmlibrary.com. SAGE Publications Ltd. Available at: http://eiilmlibrary.com/library/93342_Lesley%20Pender,%20Richard%20Sharpley%20-%20The%20Management%20of%20Tourism-Sage%20Publications%20Ltd%20(2004).pdf (accessed on 14 June 2023).
7. Thirumoorthi, T., & Mun, B. W. K., (2015). *(PDF) Tourism Theories –* ResearchGate, www.researchgate.net. UM Press. Available at: https://www.researchgate.net/publication/282124420_Tourism_Theories (accessed on 14 June 2023).
8. *Tourism Concepts and Practices,* (2016). https://cbseacademic.nic.in/. The Secretary, Central Board of Secondary Education. Available at: http://cbseacademic.nic.in/web_material/Curriculum/Vocational/2018/Tourism/XI/Tourism%20Concepts%20and%20Practices.pdf (accessed on 14 June 2023).

Chapter

3 ENTREPRENEURSHIP EDUCATION IN TOURISM AND HOSPITALITY MANAGEMENT

Contents

3.1. Introduction to Entrepreneurship Education 58
3.2. Entrepreneur and Entrepreneurial Education 62
3.3. Characteristics of Entrepreneurship Education................................... 65
3.4. Entrepreneurship Education and the New Role of Teacher 65
3.5. How Has Entrepreneurship Education Evolved? 69
3.6. Global Entrepreneurship Education and Research............................. 70
3.7. Entrepretality 72
3.8. Preparing Entrepreneurs for Hospitality and Tourism Industry............ 74
3.9. Theoretical Grounds for Tourism Education and Entrepreneurship..... 77
3.10. Empirical Ground for Hospitality and Tourism Industry 79
3.11. Future Development of Hospitality and Tourism Education in Asia.. 80
3.12. Integration of Hotel Management Education and Innovative Entrepreneurship Education in Colleges 82
3.13. Conclusion 84
References 85

This chapter is all about entrepreneurship education in Tourism and Hospitality Management. At the beginning of this chapter, this chapter explains entrepreneurship education and entrepreneur. This chapter also explains the characteristics of entrepreneurship education and the new role of teachers in entrepreneurship education. In the middle of this chapter, it explains global entrepreneurship education and entrepretality.

The theoretical and empirical grounds for the hospitality and tourism industry will also discuss in this chapter. At the end of this chapter, this chapter also provides highlights on the future of hospitality and tourism education in Asia and on the integrations of hotel management education and innovative entrepreneurship education in colleges.

3.1. INTRODUCTION TO ENTREPRENEURSHIP EDUCATION

There is a lot of ambiguity surrounding the term "entrepreneurship" due to its significant variation between businesses and academic fields. "The ambition, drive, and abilities necessary to establish and operate a successful firm," according to Sewell and Pool, characterize entrepreneurship. According to Hansen, Shrader, and Monllor (2011), how one defines opportunity affects how one defines entrepreneurship.

Kobia and Sikalieh (2010) write that "a clear-cut and controversy-free definition is nowhere within grasp." Additionally, innovation and entrepreneurship are frequently used interchangeably, yet there is a significant difference between the two.

The different ways that entrepreneurship is defined in education make it difficult to develop entrepreneurship curricula both inside schools and departments and when discussing entrepreneurship across academic disciplines. Therefore, in this chapter, we frame the problem using a foundational purpose rather than an operational definition.

Encourage students to start their own enterprises and help them acquire the knowledge and skills they need to succeed in their own endeavors is a widely regarded fundamental goal of entrepreneurship education (Harkema & Schout, 2008).

For the purposes of this chapter, we recommend an improved working definition of entrepreneurship education. With the addition of this new goal, the working definition of entrepreneurship now more accurately describes how it applies to the hotel sector. According to this extra goal, students are also urged to innovate inside already established companies in order to offer value.

This additional goal may be described as innovation or corporate entrepreneurship. However, the definition of corporate entrepreneurship provided by Burns (2005) only refers to adding value to already-existing, sizable organizations. The objective we suggest does not take the organization's overall size into account when determining where value is being created.

The organization could be large or small. Additionally, neither the ownership nor the aim of the company are taken into account by the goal we propose. For instance, the business might be a corporation or a partnership; it might also be for profit or not for profit. Utilizing resources that are already accessible to produce something of worth is the extra goal, which is consistent with value creation.

The idea of entrepreneurship is elusive, hard to define, and has many diverse interpretations because different academics have varied perspectives on it depending on the circumstances (Figure 3.1). For instance, the psychologist's (behavioralist's) definition of it is "the urge for achievement, perceived locus of control, and risk-taking propensity."

Figure 3.1. Reshaping the World through Entrepreneurship, Education and Employment: Klaus Schwab, Park Geun-Hye.

Soure: Image by Flickr

The economist sees it as combining the factors of production (land, labor, capital, and entrepreneur) and taking on the risk of both purchasing at a fixed price and selling at uncertain prices. The sociologist, however, sees it as the capacity to spot and seize market possibilities in order to deliver social services. Since each of these strategies concentrates on only a portion of entrepreneurship while ignoring others, neither is sound or comprehensive. However, four distinct definitions of the term "entrepreneurship" provided by various academics are listed below:

Education in entrepreneurship refers to a person's attitude and capacity to look for investment opportunities in a setting and to successfully launch and manage a business.

Entrepreneurship education is defined as the process of producing something unique and valuable while investing the necessary time and effort, taking on the associated financial, psychological, and social risk, and reaping the financial and personal satisfaction.

Education in entrepreneurship is seen as an effort to produce value by identifying business possibilities and developing the management, communication, and other skills needed to mobilize the people, financial, and material resources required to complete a project.

Identification, development, and realization of a vision are all steps in the entrepreneurial education process. The vision might be an innovative idea, an opportunity, or just a better way to carry out a task. This procedure results in the formation of a new venture that was founded in a risky and unpredictable environment.

An entrepreneur, on the other hand, is a person who decides to take on the financial risk of starting and running a new business. The business venture may be based on an entirely original concept, a novel method of operation, a brand-new setting, or an attempt at something that has never been done before. To put it another way, an entrepreneur is viewed as someone who recognizes an unrealized potential for significant profits (either by reducing the cost of manufacturing already-existing goods or services or by developing entirely new ones).

A definition serves as a springboard for a thorough comprehension of the phenomenon under consideration. One might be able to learn about the essence, issues, and goals of the phenomenon through it as well. Similarly, one can speculatively discuss the essence, concerns, and goals of entrepreneurship as a subject of study (Hytti, 2002; Hytti and O'Gorman, 2004; Jones and English, 2004; Henry et al., 2005a, b; Gartner, 1990).

These fundamental concerns, which arise from a definition, may later be used as the foundation for conceptually matching entrepreneurship education with the right target audience, course materials, and teaching approaches.

However, in order to make things more convenient, the next paragraph will express the views of certain reviewers in the field. Conflicting entrepreneurial schools of thought and a general lack of a definition of entrepreneurship have been discovered by evaluating the material already published (Sexton and Bowman, 1984) (Figure 3.2).

Figure 3.2. Education Development Centre, Enterprise Way, Spennymoor.

Source: Image by Wikimedia Commons

There is disagreement over the use of terms like "entrepreneurship education" and "enterprise education," as well as the substitution of "entrepreneurship education" with "entrepreneurial education." Advocates claim that there is a conceptual difference between the two: the former has to do with developing an attitude of independence, while the latter is for developing people who look for opportunities.

However, to others, although having different contexts, the two phrases are semantically equivalent. According to Gibb (1993), referenced in Fank et al. (2005), the terms "entrepreneurship education" and "enterprise education" are mostly used in the United States and Canada.

As a result, entrepreneurship education can be described as the process of applying knowledge, attitude, skills, and competencies in a professional setting. It goes beyond instructing students in how to run their own independent businesses. It involves cultivating a climate in the classroom that encourages entrepreneurial behaviors and attitudes, such as taking calculated risks, being responsible, and appreciating diversity.

3.2. ENTREPRENEUR AND ENTREPRENEURIAL EDUCATION

Entrepreneurship can be defined as the ability to offer answers and develop a system targeted at improving productivity in various industrial and governmental domains (Aggarua & Esposito, 2001). In other words, a person who can plan, manage, and take measured risks in an economic activity is considered an entrepreneur.

According to Shane and Ulrich (2001), entrepreneurship in the management sciences emphasizes four factors: risk-taking, organizational design, strategy and performance, and decision-making.

According to Aisrich and Peters and the literature, the idea of entrepreneurship alludes to a time in the future when conventional management theories will no longer apply and will be replaced by new ones.

Aisrich and Peters (2005) went on to add that entrepreneurship is a process through which something novel with nearly new values is accomplished while investing acceptable time and effort, considering the financial and social risks, and achieving monetary benefits and job satisfaction.

According to Brookes and Altinay (2015), Schumpeter (1934: 354) is recognized as one of the pioneering academics who characterized entrepreneurs. Entrepreneurs are those, according to Schumpeter, "whose role it is to carry out innovative combinations of means of production." Spencer, Kirchnoff, and White (2008) claim there is no single definition of entrepreneurship despite the fact that it is seen as the method by which people perform this duty.

This is because entrepreneurship is described from many different angles. The definition of an entrepreneur includes not only someone who can take risks and launch a firm, but also someone who uses their abilities and unique personalities to add value to organizations (Gundry, Ofstein & Kickul, 2014).

Entrepreneurship is not just about starting new businesses; it can also happen in established ones (Shane & Venkataraman, 2007). Additionally, it encompasses entrepreneurial prospects as well as the relationships between the individual and the opportunity, or the "individual-opportunity" as described by Shane (2003).

According to Stevenson and Jarillo (1990) Entrepreneurship is "a process through which individuals – either on their own or inside of organizations – utilize possibilities without regard to the resources they already possess,"

According to Lundstrom et al. (2012), The European Council (2006) identified eight critical competencies that everyone should possess to facilitate the establishment and innovation of new businesses. This viewpoint is generally held because young university graduates today have a high unemployment rate and few work opportunities in business enterprises.

Additionally, the current generation of young graduates does not find paid employment in companies to be attractive; instead, they see achievement, flexibility, and creativity as major workplace motivators. Additionally, barriers to entrepreneurship are easily overcome.

The majority of recent graduates are constantly looking forward to either starting up fast-growing businesses of their own or working in large corporations, feeling that this offers a brighter future. Someone who possesses entrepreneurial skills can lead a successful working life (Daniel, Costa, Pita & Costa, 2017).

Entrepreneurship in the travel and tourism industry has been cited as a key driver of the growth of local communities' social fabric. Responsible entrepreneurship is not just about resolving economic issues. According to Matthias et al. (2019), responsible enterprise can also address social issues like enhancing social cohesion and lowering inequality. It aims to establish equity and a balance between economic growth, social demands, and environmental goals. The aforementioned have prompted the tertiary institutions to take into consideration developing tourism education programs that lay emphasis on teaching and improving individual's entrepreneurial skills while taking cognizance of the fact that new graduates may prefer to become self-employed or innovative employees and enjoy a successful professional life (Daniel et al., 2017), and also become self-reliant (Figure 3.3).

Education gives business owners the non-cognitive and cognitive competencies they need to better assess, evaluate, spot, and take advantage of entrepreneurial possibilities. It also boosts their level of confidence and lowers their perception of risk. According to Dershuis Van Justion (2003) and Vakili et al. (2016), education is one of the main tools for cultivating an entrepreneurial spirit.

Figure 3.3. *YSEALI Cambodia – eco-tourism and entrepreneurship for all.*

Source: Image by Flickr

According to Lestari et al. (2018), the Danish Foundation for Entrepreneurship has defined entrepreneurial education as "Content, methods, and activities supporting the creation of knowledge, competencies, and experiences that enable students to initiate and participate in entrepreneurial value creation processes" (Moberg, Sternberg, & Vestergaad, 2012, p. 14).

This concept of entrepreneurial education is based on the statement that "entrepreneurship is the act of seizing opportunities and ideas and re-creating them into value for others. The value that is created can be financial, cultural, or social" (Moberg et al., 2012, p.14).

The idea of entrepreneurial value creation, or the idea that the value created should be innovative but also that it involves some kind of initiative on the side of the person who is the value creator, is something that is common to definitions of entrepreneurs., the value creation process is managed and owned by the process's starter (the student), who also carries the risk of failure, and that it entails acquiring the resources necessary to create the value (Shapero & Sokol, 1982; Okpara & Halkias, 2011).

Students' aspirations to start their own businesses are positively impacted by university programs that promote entrepreneurship.

3.3. CHARACTERISTICS OF ENTREPRENEURSHIP EDUCATION

The development of entrepreneurial culture is essentially the aim of entrepreneurship education. It aids prospective business owners in locating and seizing opportunities. It goes beyond encouraging start-ups, creative businesses, and new jobs.

All people can be entrepreneurial, which helps young people develop their creativity and self-confidence in all of their endeavors. The authors have determined the following to be the fundamental characteristics of entrepreneurship education as a field based on a careful analysis of the relevant literature:

1. It is a function of innovation.
2. It is a function of fostering leadership.
3. It is an organizational building function.
4. It is a function of high achievement.
5. It involves creation and operation of an enterprise.
6. It is process of creating value for customers by exploiting untapped opportunities.
7. It is strong and positive orientation towards growth in wealth, knowledge and employment.
8. It is concerned with attitudinal change, risk taking abilities and turning idea in to actions.

Therefore, entrepreneurship education as a discipline always strives to instill some skills, so one can play a role as a catalyst for socioeconomic change. It provides motivation to simultaneously influence one's personal life and the society of the future.

3.4. ENTREPRENEURSHIP EDUCATION AND THE NEW ROLE OF TEACHER

All activities intended to promote entrepreneurial mindsets, attitudes, and abilities fall under the category of "entrepreneurship education," which encompasses a wide variety of topics such idea development, start-up, growth, and innovation (Fayolle, 2009).

According to Mwasalwiba (2010), Shigeru Fijii, who began teaching in this area in 1938 at Kobe University in Japan, was a pioneer in the subject of entrepreneurship education. The 1940s saw the emergence of small business management courses. At Harvard Business School, Myles Mace introduced the first entrepreneurship course in the country in 1947.

This phenomena were only acknowledged by the general public in the last 50 years (Alberti et al., 2004). Nearly all American Assembly of College Schools of Business (AACSB) recognized institutions offer entrepreneurship courses, totaling approximately 1400 postsecondary institutions.

Many people have expressed opinions on whether entrepreneurial goals can be attained and strengthened through education and training or whether some people are "born" to be entrepreneurs or act in an entrepreneurial manner in the midst of the continuous increase in the number of universities offering entrepreneurship courses. Some continue to insist that entrepreneurship cannot be taught. For them, personality and psychological traits are important factors in entrepreneurship. One of the arguments put out is that aptitude and temperament cannot be taught (Fayolle et al, 2008; Fayolle, 2007; Thompson, 2004). However, one may contend that this is true for a variety of vocations and workplace settings. There is no denying that teaching abilities may be taught in the fields of medicine, law, engineering, and education, yet some doctors, attorneys, engineers, and teachers are brilliant while others are not. Similar justifications can be offered for entrepreneurship and entrepreneurs (Fayolle et al, 2008; Hindel, 2004; Fayolle, 2007). According to Shepherd and Douglas (1997) and Rae and Carswell (2001), there is a difference between the teachable and unteachable aspects of entrepreneurship.

Figure 3.4. XLRI Campus.

Source: Image by Wikimedia Commons

The best strategy to manage teachable skills and establish the right balance between student needs and instructional strategies are essential for a successful entrepreneurial education (Lee et al., 2007). Mwasalwiba (2010) stated that it is without question that young entrepreneurs can be taught the fundamental entrepreneurial skill, knowledge, and attitude notwithstanding the ongoing discussions on the teaching and training components of entrepreneurship (Figure 3.4).

Jamieson (1984, p. 19), referenced in Faoite et al. (2003), defined entrepreneurship education as a discipline as "the training of ability, knowledge and attitude for people to go out and produce their own returns and solve their issues." Scatt et al. (1998) & Matlay and Mitra (2002) categorized entrepreneurship education in following three different ways:

1. Education 'about' enterprise (Awareness Creation);
2. Education 'for' enterprise (The preparation of aspiring entrepreneurs for innovation); and
3. Education 'in' enterprise (The growth and development training for established entrepreneurs).

Jamieson (1984), quoted by Faoite et al., proposed a similar categorization (2003). Additionally, they describe this type of education as being "about" business (as a significant force for social and economic change), "through" business (teaching methods that incorporate entrepreneurial situations), and "for" business (training both potential and existing entrepreneurs).

Therefore, entrepreneurship education as a discipline holds that entrepreneurs are not born but rather develop via their life experiences. Entrepreneurship education is a dynamic process of vision, transformation, and invention, claim Kurtako and Hogges (2004).

According to Johannisson and Olaison (2006), two perspectives on entrepreneurship are frequently used: entrepreneurship as a tool or instrument for management and entrepreneurship as forms of social creativity (Johannisson, 1992; Hjorth, 2003; Dey & Steyaert, 2006, as cited in Ruskovaara & Ikävalko, 2007).

According to The Consortium for Entrepreneurship Education (2008), entrepreneurship education is not just about educating people how to run a business; it also aims to foster their creative thinking and promote their entrepreneurial spirit. It takes more than just acquiring knowledge to establish entrepreneurship as a vital talent. Since entrepreneurship education focuses on fostering the ability to behave entrepreneurially, attitude and behavior may be more significant than understanding on how to operate a business. According to the European Commission (2011), creating a culture that is through, for, and about entrepreneurship constitutes entrepreneurship education.

The greatest way for children to develop these skills is through inquiry-driven, discovery-driven learning that empowers them to put their ideas into action. They are challenging to teach using conventional methods of teaching and learning, in which the pupil is frequently a more or less passive recipient (Figure 3.5).

Figure 3.5. Entrepreneurship Development Institute of India (EDI).

Source: Image by Wikimedia Commons

They need dynamic, learner-centered pedagogies and instructional strategies that draw on real-world, applicable learning opportunities. Further, as entrepreneurship education is a transversal competence, it ought to be made available to all students and taught as a theme rather than as a separate subject at all levels and stages of education (ECOTEC, 2010).

Consequently, the role of instructors must shift for young people to achieve above-average entrepreneurial competencies (European commission,

2004). Entrepreneurship education requires a fundamental shift in the way teachers educate, shifting from being knowledge communicators to learning facilitators (European Commission, 2013; World Economic Forum, 2009). There is no doubt that these changes have substantial implications for instructors.

They imply that all teachers will now play the role of "learning facilitators," which is nothing less. At a point where many change processes connected to entrepreneurship education converge, teachers are in a precarious position (Hannon, 2006; Hytti and O'Gorman, 2004). At times, they have struggled to decide how to react to the problems that are present. Jaana (2008) as referenced in Ruskovaara & Ikävalko's (2007) work; (Fiet, 2001).

All of this indicates that there is a lot of interest in effective entrepreneurship education; expectations are high, and teachers must play a key role in helping expectations be met. All of this indicates that there is a lot of interest in effective entrepreneurship education; expectations are high, and teachers must play a key role in helping expectations be met.

3.5. HOW HAS ENTREPRENEURSHIP EDUCATION EVOLVED?

Can you teach entrepreneurship is a persistent conundrum at many colleges? Many years before the Industrial Revolution, there were general definitions of entrepreneurship (Matlay, 2005). However, since the 1940s, entrepreneurship education has been legally acknowledged at institutions. The primary focus of undergraduate courses has not hitherto been entrepreneurship education.

The strategic development of curriculum across programs has turned its attention to entrepreneurship education during the past ten years (Katz, 2003). In spite of the fact that "employability, enterprise, and entrepreneurship are high on the agendas" of institutions of higher education, Sewell and Pool (2010) note that it is challenging to align strategic plans with curriculum development because there is no consensus on what exactly constitutes entrepreneurship.

The focus of entrepreneurial education has typically been at business institutions. Business schools are quite skilled at showcasing to students, graduates, professors, and the neighborhood the direct economic impact of their activities (Katz, 2003). The ability to articulate the value of an education is a prerequisite for the creation and continuous support of business school programs. The relationship between these programs and the creation of economic value has supported and encouraged the growth of entrepreneurship education programs in the for-profit sector. Administrators have started to focus more on where entrepreneurship education should be anchored inside a University since upgrading their long-term strategic plans to remain competitive on a national and worldwide level.

Administrators of universities and economic development officials prioritize efficiency in light of the challenges of an increasingly competitive global environment. Through entrepreneurship education, the objective is to promote the development of businesses that are both profit- and socially-driven.

They are trying to figure out the optimum method for incorporating different academic disciplines into entrepreneurial curriculum. In the future, it is hoped that entrepreneurship education programs would be designed to best meet the needs of a varied undergraduate student population with a range of disciplinary interests.

Despite their shared origins, liberal arts and entrepreneurship are seen as being on opposite sides of the academy by critics of entrepreneurship education because these programs produce profit-driven, commercial firms (Godwyn, 2009). Notably, business programs have led for-profit entrepreneurial education while liberal arts studies have been at the forefront of social entrepreneurship activities.

This emphasis has brought attention to the necessity for institutions to get rid of their stovepipe mentalities. Through entrepreneurial education, there is a trend to promote the establishment of both profit-driven and socially-driven enterprises, which is in line with the growing trend to identify efficiencies in multidisciplinary education (Pribadi, 2005). Innovation has been hindered by most universities' "stovepipe" structures.

3.6. GLOBAL ENTREPRENEURSHIP EDUCATION AND RESEARCH

The global influence of entrepreneurship is increasing. Improvements in the production of our commodities and innovation in service delivery are both products of entrepreneurship. Within universities, there is a growing demand for entrepreneurial education at different levels. More entrepreneurship courses are being taught by faculty members.Because they were exposed to entrepreneurship while in school, undergraduates who go on to get degrees and become alumni are more likely to start their own small enterprises or innovate inside already established ones. Entrepreneurs are persons who create valuable products based on opportunities they recognize and the resources they have access to in a particular area.

Entrepreneurship is not a new topic for research. The concept of "entrepreneurship education" is constantly changing, and there are still a lot of undiscovered research possibilities. The question of whether entrepreneurship education is teachable or not, and if it is, what kind of model should be used, is one that frequently appears in papers and research studies (Kleiman, 2008).

Ironically, the models and data presented demonstrate how difficult it is to define "entrepreneurship" because it is a term that is constantly growing (Pribadi, 2005).

Before choosing a model to construct the curriculum for a university as a whole, Pribadi (2005) points out that the integration of entrepreneurship into particular ecosystems, such as different university departments, needs to be researched.

Infrastructure and institutions are crucial in determining how programs are rewarded in a system (Soskice, 1994). To successfully integrate entrepreneurial education into the curriculum, it is essential to understand the culture and history of universities and the departments that make up those universities. This is especially true given that the ability to recognize and document the value of entrepreneurship education has not yet been thoroughly studied.

Whatever method is used, entrepreneurship education keeps expanding on a global scale. According to Katz's article from 2003, 120,000 American students were enrolled in entrepreneurship courses in 1994, and by the start of the new century, that number had climbed by 50%. In recent years, entrepreneurship education has experienced some of its most rapid growth. This growth may have exceeded the capacity of the existing intellectual resources.

Furthermore, Katz (2003) observes that the capacity to attract the smartest and brightest students to study at American colleges is significantly impacted by worldwide competition. This means that American business school entrepreneurship professors will face increased competition for the best students, innovative ideas, and prestigious publishing venues from entrepreneurship faculty members not only around the world but also from entrepreneurship faculty members in the office next door.

The assessment techniques used for these programs make it challenging to analyze national and international growth in entrepreneurship education. According to Lee (2005), it is more relevant to gauge the socioeconomic impact that entrepreneurship programs have on the enterprises that are created than it is to gauge the number of students who graduate from them.

It is crucial to track economic growth using indicators like the number of businesses founded, the number of jobs produced, the sorts of businesses formed, and the potential for future expansion of those businesses.

3.7. ENTREPRETALITY

The goal of entrepreneurship education in the hospitality business is described by a new term we coined, "entrepretality." The concepts "entrepreneurship" and "hospitality" are combined literally to form the word. By expanding the service-value proposition to a client, customer, or other being, entrepreneurs generate value by utilizing the resources at their disposal.

The definition of entrepreneurship, which places a strong emphasis on the worth of a service output, differs significantly from the notion of entrepreneurship that is most frequently accepted in the academic community of hospitality.

According to Brizek and Khan (2008), "entrepreneurship is mostly considered in terms of individuals who founded new businesses" in the academic field of hospitality. This definition of entrepreneurship in the hospitality academic community is limiting since it only looks at the establishment of new businesses and ignores the foundation of hospitality education, such as developing a service-value proposition for a client.

The need to connect the theoretical with the practical in hospitality education is constant and has several facets (Naipaul & Youcheng, 2009). Therefore, hospitality programs are a great fit to combine theory and experience in order to look at boosting entrepreneurship.

Almost all hospitality courses can include entrepreneurial concepts. For instance, a course in restaurant or hotel design might promote entrepreneurship. A professor may emphasize the significance of a higher service level in one or both of these courses. There are various strategies for incorporating entrepreneurialism at the industry level.

Entrepreneurship courses, concentrations, and practicums are now being offered by reputable hospitality colleges all around the world. Additionally, business proposal and business model competitions are held in hospitality programs, and mentorship exchanges between students and alumni are encouraged. Cornell University is a prime example of entrepreneurialism. The Leland C. and Mary M. Pillsbury Institute for Hotel Entrepreneurship were established in 2006.

The School of Hotel Administration at Cornell University now offers a number of new entrepreneurship courses as a result of the center's establishment. Since the actual situation in which the courses are presented, also known as contextual learning, is a prevalent subject among hotel schools, entrepreneurship is a particularly appealing component to take into account for a program.

3.7.1. Implications for the Future for Entrepretality

The importance of entrepreneurship will increase as the economy becomes more competitive, shifting, and global. "Intensifying global competition, corporate downsizing and de-layering and rapid technological progress have heightened the need for organizations to become more entrepreneurial to survive and prosper" (Li, Tse, & Zhao, 2009).

To adapt to the shifting demands of the global economy, university programs will need to be adaptable enough. Strategic planning will continue to place a high priority on maximizing the effectiveness of entrepreneurship education across the university.

The skill sets of graduates will need to change to meet the industry's evolving needs. According to Weiermair, Siller, and Mössenlechner (2006), the attempt to promote employability will have an impact on how innovative the hotel sector is and how entrepreneurship education is perceived.

The curriculum will change to reflect this change, and the number of courses that emphasize entrepreneurship or that provide entrepreneurship courses will increase. In order to educate future entrepreneurs, managers, and leaders who are intellectually capable of thinking outside of current business methods and paradigms, curriculum content will place a greater emphasis on self-discovery, critical, and creative thinking (Naipaul & Youcheng, 2009).

The connection between the theoretical and the practical will become increasingly crucial and critical to assuring future innovation in entrepreneurship. It might be possible to find a workable solution by incorporating instructional models from fields that use HPI and contextual learning.

Finally, the desire to promote innovation within an existing business, regardless of size, as well as entrepreneurship in launching new businesses will be evident in the expansion of multidisciplinary training.

As more entrepreneurial courses are developed, more PhD students will pursue their entrepreneurship degrees. This rise in PhD candidates will result in an immediate improvement in the standard and quantity of entrepreneurship-related research.

The fundamental question is whether entrepreneurship will be taught in a hospitality program or if it will be made available at the university level to students with a wider range of backgrounds and skill sets, regardless of academic field.

In addition, regardless of the industry, entrepreneurship is a critical set of skills for a successful businessperson. Strong interpersonal and hospitality skills are essential, and hotel programs all over the world can use this as a tool to not only attract new students but also guarantee a key spot in their institution's long-term curriculum.

3.8. PREPARING ENTREPRENEURS FOR HOSPITALITY AND TOURISM INDUSTRY

For those who wish to pursue an entrepreneurial education career, it is now customary to use an educational model that combines practical hospitality and tourism courses with management lectures and industrial training to give students real-world experience and allow them to apply what they have learned (Ciochina, Iordache, Sirbu & Alexandrina, 2016).

For instance, the curriculum in Nigeria has traditionally included Students Industrial Work Experience (SIWES). The industrial training provides students with the chance to hone their abilities. SIWES effectiveness is influenced by a variety of factors, including training intensity. There is no proof, as of the time of this assessment, that the efficiency of SIWES has been properly examined.

According to Daniel et al. (2017), the goal of entrepreneurship education is to equip students with the skills necessary to start their own initiatives and businesses. The students learn how to create their own hospitality/tourism business concepts and compete against other students across the world.

According to researchers and academics, the society we live in now is going through a period of transition, thus past experiences, methods, and fixes might not be applicable to addressing organizational problems today.

Therefore, institutionalizing the spirit of entrepreneurship in businesses and teaching it in colleges and universities are necessary to create entrepreneurs who can compete successfully on a global scale. (Kuratko & Hodgetts, 2001).

Through entrepreneurial education, students acquire the attitudes, skills, and behaviors needed to deal with obstacles in life. The idea of entrepreneurial knowledge has changed and evolved over the last few years.

According to Lestari et al. (2012:1), entrepreneurship education can help students develop the mindset, attitude, and behavior of a true entrepreneur and can even influence them to choose entrepreneurship as a career. In Lestari et al., the entrepreneurial learning process imparts information and skills in addition to the capacity to pursue and achieve one's own objectives in the context of business (Siswadi, 2013).

Universities are expected to play a significant role in the education of tourism professionals; as a result, how well they fulfill this role by fostering an environment that is conducive to teaching and learning, among other things, will greatly influence the effectiveness of entrepreneurship education Ntshangase & Ezeuduji (2019).

According to Brown (2000), who is quoted by Izedonmi and Okafor (2010:10), entrepreneurial education curricula should be organized to convey and instill the competences, abilities, and values necessary to recognize business opportunities, seize them, organize, and launch new company enterprises.

According to Johnson (1988), Heinonen and Hytti (2010), and O'Connor (2013), there are three main categories of entrepreneurial education: Teaching "about" entrepreneurship refers to a theoretical and content-heavy approach designed to provide a broad grasp of the phenomenon.

The majority of higher education institutions use this strategy (Mwasalwiba, 2010). Occupationally focused instruction designed to equip aspiring business owners with the necessary knowledge and skills is referred to as "teaching for" entrepreneurship. According to Kyrö (2005), teaching "through" refers to a method that is process-based and frequently experiential and in which students actually engage in an entrepreneurial learning process.

The essential business, technical, and managerial abilities needed to succeed as an entrepreneur are required for those in the hotel and tourism industries Ciochina (2016). These operators oversee the day-to-day operations of the businesses, acting in managerial capacities such as planning, controlling, budgeting, making decisions, organizing, and managing all facets of the tourism industry.

They also keep an eye on the accounting and cash/financial control procedures to make sure that organizational objectives are met. The operators must therefore receive appropriate training in entrepreneurship. Morris, Nebb, and Singhal (2013) and Sanchez (2013) emphasize the significance of fostering students' analytical, contrasting, and evaluating thinking skills as well as their imaginative and exploratory abilities.

Entrepreneurship education, according to Cooper, Bottomley, and Gordon (2004), serves three crucial functions:

1. Strengthening the capacity to generate ideas and innovation,
2. Accelerating business application of these and innovations and extending these capacities, and
3. Establishing entrepreneurship education initiatives that increase interest in current and transferrable entrepreneurship programs among former graduates.

According to Linan (2007), Sanchez (2013), Rae (2005), Nabi and Holden (2008), Saarinen and Ursin (20012), and Gibb (2005), entrepreneurial education programs should integrate curricular (theoretically and organized education) and extracurricular activities.

This is in response to the ongoing debate regarding the relationship between higher education institutions and entrepreneurship. Further, these authors advocate for the idea that extracurricular activities have a favorable effect on a career as an entrepreneur.

Extracurricular activities include learning-by-doing, publishing periodicals and materials generally, visiting companies and giving seminars on entrepreneurship by outside speakers, as well as other activities that communicate the university's purpose and commitment to the values of entrepreneurship.

Universities should play a crucial role in helping students develop their entrepreneurial vocations, according to Collins, Hannon, and Smith (2004), Ertuna and Gurel (2011), and Lestari et al. (2018). In support of above view, Suharti (2012) in Lestari (2018), Kirby (2004), Maskell and Robinson (2002), Etzkowtz (2004), and Saarinen and Ursin (2012) have proposed the concept of an entrepreneurial university in which the pattern of a set of perfectly controlled actions with a systemic approach instills entrepreneurial behavior as a culture rather than the exception.

According to Kirby's (2005) theory, an entrepreneurial university has the ability to innovate, spot and seize environmental chances, foster teamwork, take calculated risks, and respond to problems with solutions.

3.9. THEORETICAL GROUNDS FOR TOURISM EDUCATION AND ENTREPRENEURSHIP

According to Ajzen (1991); Shapero and Sokol (1982) in (Ajzen, 1991); Kolvereid (1996); Krueger (2000); and Shapero & Sokol (1982) in Arranz et al., intention is the driving force behind behavior and is viewed as its immediate antecedent.

Three factors play a role in the development of an intention: a positive or negative assessment of the activity; perceptions about the wishes of significant reference groups to perform or refrain from performing the action; and the perceived ability to perform the behavior. (Krueger, 2000).

Researchers and scholars have questioned whether the entrepreneurial training programs provided to university students majoring in hospitality and tourism have changed their behavior and provided the necessary competence (knowledge, skill and attitude).

Shapero and Sokol (1982) and Ajzen (1991) offer a theoretical framework that divides activities involved in higher education into "curricular" and "extracurricular" categories in order to address this dilemma. This framework is founded on the psychological underpinnings of intentional theory. While extracurricular activities assist cognitive, informative/formative, and instrumental activities, curricular activities require involvement in formal learning schemes that involve the teaching of skills and attitudes.

Similar to this, Sanchez (2013) examined the outcomes of entrepreneurship courses and reported that the creation of educational programs had an impact on entrepreneurship-related abilities including taking risks, having self-confidence, and being proactive.

A multidisciplinary approach is taken to the organization of education in the tourist and hospitality fields, combining technical training with managerial and business abilities. The majority of efforts made by higher education institutions to foster entrepreneurial vocations used a psychological strategy, which involves fostering university students' intents, attitudes, and behavioral control.

Through the investigation of perceptions and intents, the cognitive psychology approach provides a useful viewpoint on how to investigate the phenomenon of entrepreneurial decision-making. To the subject of whether the entrepreneurial training programs provided to university students majoring in hospitality and tourism have affected their behavior and taught the necessary abilities, scholars and researchers have offered varying responses (knowledge, skill, and attitude) (1982 by Shapero & Sokol; 1991 by Ajazen). While some researchers claim that entrepreneurship education has no, negative, or none of those effects on pupils, others claim that the training has a favorable effect (Sanchez, 2013). We support the Theory of Planned Behavior in our contribution to the discussion, as well as the idea that entrepreneurship education benefits students. The Expectancy Theory of Motivation by Vroom (1964), as cited in Igbojekwe and Agbonye, is a socio psychological theory that informs our perspective (2009).

Theories of motivation typically explain what inspires, guides, and sustains someone's efforts (behaviors). According to the expectancy theory of motivation (ETM), an offer's antecedents, contents, and effects on behavior are all explained.

This theory's fundamental premises are that people are driven by the value (valence) and attractiveness of the expected outcome and that expectations have an impact on people's behavior. The theory further emphasizes that there should be a connection between effort and performance (outcome), as well as a connection between reward and performance (instrumentality)

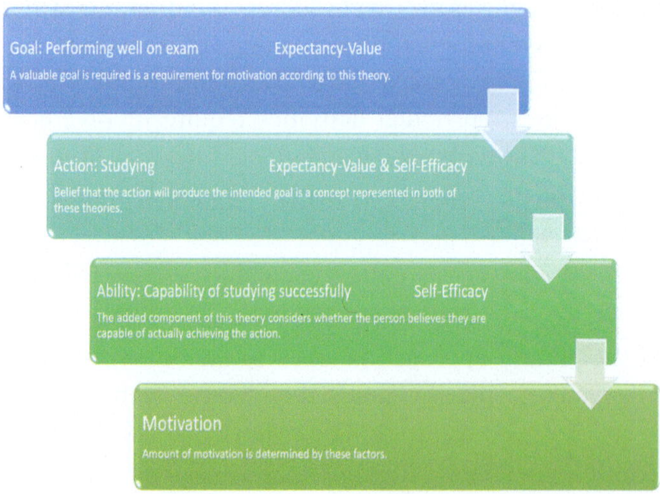

Figure 3.6. Expectancy theory of motivation (ETM).

Source: Image by Wikimedia Commons

We try to provide an explanation for the phenomenon of entrepreneurial education and its effects on students from the standpoint of ETM. The aim of the training is to get students ready to gain the skills they need to start profitable businesses once they graduate from college, which is their desired goal and the intended outcome of the tourist entrepreneurship education (Figure 3.6).

Young graduates of universities seek to work for a rapidly expanding company organization or to become successful professionals who are self-employed and independent. They would be motivated to acquire entrepreneurial education through which they build self-confidence, and by the perceived achievements and benefits derivable from being tourism entrepreneurs.

3.10. EMPIRICAL GROUND FOR HOSPITALITY AND TOURISM INDUSTRY

The relationship between entrepreneurial education for the tourism industry and its impact on entrepreneurial spirit has been the subject of investigations by researchers. However, the majority of studies (such as Autio et al. 1997; Boissin et al. 2009; and Elmuti and Khoury 2012) examined the effectiveness of initiatives to encourage entrepreneurial vocations and came to the conclusion that including subjects related to entrepreneurship in the curriculum may encourage participants' propensity to pursue it.

According to studies done with university students by Arranz et al. (2017) and Elmuti and Khoury (2012), taking part in subjects that involved starting a business showed a positive propensity to participants' entrepreneurial intention.

According to Lian and Chen (2009), Boissin et al. (2009), and Sanchez (2013), earlier research has demonstrated that a university education has a favorable impact on the acquisition of the necessary skills and competences for fostering an entrepreneurial spirit. According to Garavan and O'Cinneide (1994), entrepreneurship may be positively or negatively impacted by education.

According to Peterman and Kennedy (2003), formal education "does not inherently inspire entrepreneurial vocation, but instead leads to conformity and a reduce tolerance for uncertainty, weakens students' capacity for innovative thinking and inhibits their creativity and entrepreneurial spirit." Laukkanen (2000) emphasizes that higher education institutions prevent their students from beginning new businesses by making them highly analytical, overly risk-averse, and conscious of difficulties.

The controversy surrounding the teaching of entrepreneurial vocation as a subject has been noted by Grebel et al. (2003) in Arranz et al. (2017) and Ertuna and Gurel (2011). Entrepreneurial training should strive to increase the viability and desirability of starting a firm, according to Arranz et al. (2017) and Ajazen (1991).

According to Laukkanen (2000), entrepreneurial education should be a core component of every educational system. The pedagogy of the teacher has been identified as a critical factor to take into account while developing entrepreneurial skills and aptitudes, according to Laukkanen (2000), who observes that it has an impact on the outcomes of the educational system.

3.11. FUTURE DEVELOPMENT OF HOSPITALITY AND TOURISM EDUCATION IN ASIA

The global marketization and privatization of higher education, together with the expansion of higher education providers, have led to a much expanded diversity in the higher education ecosystem. When designing new academic programs, Asian societies are significantly more receptive to market signals than their Western counterparts because of the market's increasing influence on educational progress (Mok, 2011). This view is supported by the overwhelming volume of fresh initiatives and new programs.

In Taiwan, the government recognized the need for a skilled labor force in the hospitality and tourism industries and established the nation's first higher education institution with a hospitality focus in 1995. The college status of the academy was granted in 2000, and in 2010 it changed its name to the National Kaohsiung University of Hospitality and Tourism (National Kaohsiung University of Hospitality and Tourism, 2013).

There are currently around 130 departments in higher education institutions in Taiwan, which has a population of 23 million, that provide programs linked to the hotel and tourism industry (Hou, 2011). The number of higher education institutions, both public and private, that offer programs in hospitality and tourism has also rapidly increased in Malaysia.

According to primary statistics, there are now more than 40 such institutes. One of the thriving institutions in Malaysia that has rapidly grown in hospitality and tourism is Taylor's University, a part of the Taylor's Education Group. The institution was awarded university college status in

2006, and university status in 2010. The School of Hospitality, Tourist and Culinary Arts has grown to be Southeast Asia's largest institution for the study of hospitality and tourism management.

China's tourism industry started slowly, but its growth has been phenomenal. According to the National Tourism Administration of the People's Republic of China (2013b), there were 2,236 schools offering programs linked to tourism and hospitality at the end of 2012, 1,097 of which were higher education institutions. The remaining 1,139 were secondary vocational schools.

The curricula of Asian hospitality and tourism programs have also changed over the last few decades, moving away from being a simple duplicate of the Western curriculum and toward more of a custom design, or at the very least a localized version of the imported curriculum.

The early direct copies were primarily brought back from their Western alma maters by youthful returning faculty members. The evolution has been nourished by experience accumulation and the need to set their programs apart from other locally produced programs that adopted Western curricula and international offshore programs competing in the same market.

Numerous of these programs set themselves apart from the competition by understanding the local way of life.

There are various course or subject titles that use phrases like "Asian paradigm" and "Asian-based material." In order to assist students apply Western theories and systems in a local context, culture-based content is very frequently found in teaching materials.

The need for trained faculty members is expanding as the number of hospitality and tourism programs in higher education rises. The majority of people who teach in these programs have no formal training in hospitality or tourism. Many of them came from the fields of linguistics, gastronomy, and physical education, among others.

A further development in Asian nations is the "upgrading" of programs, whereby institutions that award lower-level diplomas now grant higher-level diplomas and higher-level diploma-granting institutions now grant degrees.

As a result, faculty members now instruct students at academic levels above those for which they were hired. The assurance of the academic rigors of those programs would be difficult if they hadn't obtained higher academic credentials to achieve at least one or two levels above the students.

3.12. INTEGRATION OF HOTEL MANAGEMENT EDUCATION AND INNOVATIVE ENTREPRENEURSHIP EDUCATION IN COLLEGES

The number of large vocational schools has gradually grown over the past several years as a result of the growth in enrollment at higher vocational schools. The number of applicants in China is gradually decreasing.

Currently, the traditional curriculum system for hotel management majors uses professional technology and skills courses as its core curriculum, cultivating and mastering the fundamental knowledge and service skills of modern hotel management, and possessing applied front-line professionals needed by the modern hotel industry.

The labor market has undergone tremendous transformation, as has the working environment, as a result of the rapid expansion of networking, intelligence, and digitalization.

In order to prepare students for the future and prepare them for change, higher vocational institutions' teaching reform efforts should not only concentrate on enhancing students' professional abilities. In the future living and work environment, we should place more emphasis on the development.

The tourism hotel sector is developing in a more positive way as people's living standards rise. The best way to accomplish this is to direct hotel management toward innovation and integration based on a knowledge of the path of future development.

One of the most recent methods to enhance vocational colleges is the merger of innovation and entrepreneurship education and professional education. However, at this time, the study and application of innovative entrepreneurship education in higher vocational colleges are primarily focused on the fields of "entrepreneurship education" and "professional education," and it is a well-known phenomenon that this type of education is alienated from majors and disciplines. The program adheres to the modern vocational education philosophy of "cultivating individuals by virtue, facing everyone, and lifelong education."

The personnel training objectives and curriculum structure are dynamically changed in response to the evolving needs of the travel and hospitality sector for complex and inventive abilities as well as the requirements of each student's particular career growth. As such, we must upgrade the hotel in order for it to maintain its place in the market competition.

Therefore, hotel management needs to be innovative in order to take on a significant role in the increasingly competitive market environment. Hotel management majors in higher vocational colleges need to follow current trends and fully integrate the professional talent training mode with the Innovative entrepreneurship concept as a foundation for training front-line employees in the hospitality industry or personnel engaged in the hotel-related service industry.

It is challenging to develop new entrepreneurial talents if we continue to use the conventional curriculum and training goals. This condition needs to be changed, and to do that, Higher Vocational Colleges' old hotel management curriculum system and teaching methods must be revised. China's hotel and tourism sectors need to develop higher vocational school systems in education. Students from higher vocational institutes who major in hotel management. There are issues with the specific implementation, and the school lacks uniform management papers.

A stepwise approach from quantitative change to qualitative change is used to cultivate students' innovative thinking and entrepreneurial skills. We can therefore obtain the mathematical model of the practice-based education method for hotel management.

For college graduates to launch their own firms, higher vocational colleges must redesign the entrepreneurship education program for the hotel management major. One of the most creative and enterprising groups are college students, who have a high need for self-employment.

A constant increase in hotel service awareness is required, and higher vocational colleges' hotel management majors should focus on raising the caliber of their students as a reserve army for developing hotels. Encourage them to develop a sense of "serving the people," firmly adhere to industry norms and standards, and exercise caution at all times.

Entrepreneurship is a practical activity that involves turning creative ideas into new products, jobs, or positions that fall under the category of behavior and action. The issue of ineffective hotel management demonstrates the focus on the effectiveness and caliber of departmental work that hotel staff members are doing, as well as their lack of a broad perspective on development. Establishing close departmental cooperation is challenging. "College graduates are not just becoming less and less referred to as job seekers, but instead, they will become entrepreneurs."

The hotel industry is a very competitive one with a sizable market, particularly the food sector, which has a sizable market and few restrictions.

Some "small and beautiful" catering models provide quick capital returns, huge or modest investment scales, and simple start-ups, making them excellent for establishing a business. Entrepreneurship and innovation are largely distinct processes.

It is a type of spirit and value orientation that dares to defy expectation and be unusual, falling under the ideological level. The hotel must use science to develop the work content of internal management and external marketing in the administration of services that are innovative and integrated.

To match customer demands, hotel hardware facilities must be scientifically adjusted through internal management innovation and integration. So, it's important to continue during this lengthy instruction. The educational model should be built with a generally uniform set of standards and with content that is relatively constant.

3.13. CONCLUSION

This chapter summarizes entrepreneurship education in tourism and hospitality management. This chapter also explains characteristics of entrepreneurship educations. The basic characteristics of entrepreneurship education as a discipline as obtained from the critical review of the works related to it have been identified by the authors as mentioned in this chapter.

This chapter also provides highlights on the new role of the teacher. Entrepreneurship is not a new topic for research. So, this chapter also explains global entrepreneurship education, and Entrepretality is a new term that we created that describes the objective of entrepreneurship education in the hospitality industry. The word is a literal blend of the two words "entrepreneurship" and "hospitality." all about the Entrepretality topic has been discussed.

Toward the end of the chapter, this chapter also explains how to prepare entrepreneur for hospitality and tourism industry has been discussed. This chapter also provides highlights on the future development and integration of hotel management education and innovative entrepreneurship education in colleges.

REFERENCES

1. Anita, K. M., (n.d). *Entrepretality: Entrepreneurship Education in the Hospitality Industry*. LibGuides. guides.library.cornell.edu. Available at: https://guides.library.cornell.edu/ecommons (accessed on 14 June 2023).
2. Gautam, M., (n.d). *Entrepreneurship Education: Concept, Characteristics and Implications for Teacher Education.* ResearchGate. Available at: https://www.researchgate.net/publication/319057540_Entrepreneurship_Education_Concept_Characteristics_and_Implications_for_Teacher_Education (accessed on 14 June 2023).
3. Shuaibu, H., et al., (2021). The concept of entrepreneurship. *Education at the Intersection of Globalization and Technology* [Preprint]. Available at: https://doi.org/10.5772/intechopen.94857.
4. *Therapist Quality as an Influence of Spa Choice Decisions Amongst* (n.d). arcjournals. Available at: https://www.arcjournals.org/pdfs/ijrth/v7-i2/3.pdf (accessed on 14 June 2023).
5. Xing, Y., (n.d). Integration of hotel management education and innovative entrepreneurship education in colleges and universities. *International Journal of New Developments in Education.* Francis Academic Press. Available at: https://francis-press.com/papers/4340 (accessed on 14 June 2023).

Chapter

4 ISSUES AND CHALLENGES IN TOURISM AND HOSPITALITY EDUCATION

Contents

4.1. Introduction 88

4.2. Sustainable Tourism 90

4.3. Tourism and Hospitality Issues Induced By Globalization 92

4.4. Developing Trends in Hospitality Education – New-Fangled Encounters 94

4.5. Getting Equipped the Future Dynamism of Hospitality Manpower 94

4.6. Present Scenario of Hospitality Education System 95

4.7. Factors Affecting Tourism Industry and Its Impacts on Global Economy of the World 96

4.8. Tourism Training and Education in Less-Developed Countries 97

4.9. Emerging Trends in Hospitality Education: New Challenges 101

4.10. Accommodation Sector 102

4.11. Conclusion 105

References 107

This chapter talks about the issues and challenges in tourism and hospitality education. At the beginning of this chapter, this chapter explains sustainable tourism and the issues in tourism and hospitality induced by globalization.

This chapter also explains the developing trends in hospitality education. This chapter explains the factors that affect the tourism industry and its impact on the global economy of the world. At the end of this chapter, this chapter provides highlights on the tourism training and education of less developing countries and emerging trends in hospitality education.

4.1. INTRODUCTION

The tourism sector contributes significantly to world development and is both positively and negatively associated with the expansion of the global economy. Due to its significance, the current investigation has been planned since March 23, 2020. The universe of the study was the world.

The main objective of the study was to evaluate the variables that have both good and negative effects on the tourist sector, as well as their effects on the world economy. Secondary data were used, and a total of 18 articles were retrieved from the internet, read several times, and the situation was examined.

According to the analysis, there is a strong correlation between the global economy's economic growth and the tourism industry, despite the fact that there are numerous factors that have both positive and negative effects on the industry and that have a long-term or short-term impact on the world's economy.

The well-known factors that both positively and negatively affect the global tourism industry include culture, peace, security, the development of the world's infrastructure, visa conveniences, natural beauty, attitudes of the populace, tourist numbers, quarantines, and the world's population.

Other factors include education, income level, price levels of various goods around the world, different languages, and hotel rates. Similar to how these factors have an impact on the global economy both favorably and adversely, they also have an impact on the tourism industry both positively and badly in the short- and long-term.

The corona virus recently spread throughout the world due to poor quarantine, closing all economic sectors around the globe. The government then forbade people from having social interactions with one another globally and instructed them to stay at home. By using this strategy, China was able to avoid the virus pandemic.

Therefore, tourism is connected to every economic sector. The world's governments make more money. The people enjoy their time while being taxed by the government very much, which increases the government's revenue and creates new jobs in various nations across the world, which enhances the socioeconomic situation of the global community and brings wealth to those nations.

There are numerous issues that negatively impact this business, however only a handful are discussed here for clarification: There are fewer tourists traveling the world as a result of terrorism, as well as issues with visas and flights. The tourism sector is also badly impacted by poor transportation; High hotel and transportation costs also lead to fewer travelers worldwide; unfriendly nations also have a detrimental impact on the global tourism sector.

The study's recommendations include that the world's top tourist destinations should have strict security; the cost of hotels and transportation worldwide should be reduced; All tourists should have access to simple visas and transportation; To all tourists from around the world, clean meals should be offered at the destination; At the global entry and exit points, there should be a good and strict quarantine; In order to avoid offending visitors, tourists should always act politely; Always respect travelers from around the globe when you are at a tourist attraction.

The tourism sector should hire trustworthy employees to manage the system. All governments should create good policies to promote tourism.

The tourism and hospitality industry is undergoing rapid change. Since tourism is not a single field but is intertwined with many other spheres of life, it is continually evolving. Consider for a moment what would need to occur for someone from Hong Kong, let's call her Ms. Wong, to travel to, say, New York, and you will see how many variables are at play.

Ms. Wong must first be motivated to travel outside of Hong Kong. She must have a reason for traveling in order to accomplish this. Why should she fly out of Hong Kong and visit somewhere else in the world? What does she need there that she can't get at home? Why should she go to New York if she is primarily interested in shopping when there are thousands of stores in Hong Kong? She needs to save up enough money to cover the cost of her flight and lodging while abroad, as well as amass enough paid vacation days to be able to leave her job, in order to be able to afford to travel. She will need to make travel and lodging arrangements by plane, will need to obtain an entrance visa in order to enter the country, and will need to take the season of her stay into account (would it be winter or summer when she arrives in New York?).In addition to politics, religion, agriculture, economics, environment, health, finance, transit, society, immigration, and education, to name a few, tourism also has connections to many other fields of study. All of these areas of human endeavor are dynamic, and adjustments made in one area will have an effect on adjustments made in other areas, which will likewise have an effect on tourism. You can use the Severe Acquired Respiratory Syndrome (SARS) outbreak in 2003 and the effects it brought about to Hong Kong tourism as a concrete example. Visitors to Hong Kong avoided the city because of concern that they would contract the disease.

Our luxury accommodations, some of the best in the world, were virtually empty; arriving flights carried only a small number of passengers; Guangdong Province farmers were unable to sell vegetables to Hong Kong hotels; and staff members of tourism-related businesses either worked fewer hours or lost their jobs.

This serves as a reminder that the demand for tourism-related goods can change abruptly, often over the course of a single night, and that the sector is highly responsive to shifts and trends in the environment in which it operates. In order to make the necessary modifications to their businesses in light of these quick-moving developments, travel agencies must be constantly vigilant in order to spot emerging trends, difficulties, and changes. This involves being equipped to deal with unforeseen events and modifications to their working environment. One approach to do this is to keep an eye on media trends that might affect the company and to set aside funds to deal with unanticipated events that are beyond the organization's control.

4.2. SUSTAINABLE TOURISM

4.2.1. The Concept of Sustainable Tourism

Sustainable development is still most frequently defined as "a process to meet the demands of the present without compromising the ability of future generations to meet their own needs," as stated in the World Commission on Environment and Development report from 1987.'

Therefore, sustainable development aims to improve everyone's quality of life in ways that are just as viable in the future as they are today. In other words, sustainable development is built on ethical standards for managing the world's resources, as well as equality in how they are used and how the advantages that result from doing so are distributed. Since its initial definition in 1987, the idea has undergone significant evolution, most notably thanks to Agenda 21, the action plan that resulted from the 1992 UN Conference on Environment and Development in Rio de Janeiro, and the implementation plan from the World Summit on Sustainable Development (Johannesburg, 2002).

Three dimensions or 'pillars' of sustainable development are now recognized and underlined. These are:

- Economic sustainability, which means generating prosperity at different levels of society and addressing the cost effectiveness of all economic activity. Crucially, it is about the viability of enterprises and activities and their ability to be maintained in the long term.
- Social sustainability, which means respecting human rights and equal opportunities for all in society. It requires an equitable distribution of benefits, with a focus on alleviating poverty. There is an emphasis on local communities, maintaining and strengthening their life support systems, recognizing and respecting different cultures and avoiding any form of exploitation.
- Environmental sustainability, which means conserving and managing resources, especially those that are not renewable or are precious in terms of life support. It requires action to minimize pollution of air, land and water, and to conserve biological diversity and natural heritage.

Figure 4.1. Randy Durband, CEO Global Sustainable Tourism Council.

Source: Image by Wikimedia Commons

It is crucial to recognize the interdependence of these three pillars, which can either strengthen one another or compete with one another. Finding a balance between both is necessary to deliver sustainable development (Figure 4.1).

4.3. TOURISM AND HOSPITALITY ISSUES INDUCED BY GLOBALIZATION

4.3.1. Globalization and Its Effects on Tourism Development

"In essence, globalization is the process by which a network of relationships that span national political boundaries and unite disparate communities into a single, interdependent whole. Local differences are gradually lost as a result of this process and are absorbed into a vast global social order (Mowforth and Mundt 1998:12). It is made possible by the quick international flow of people, information, money, and ideas.

How is globalization felt in the tourism context? We can think of several ways which include the following:

- Commodification and trivialization of local cultures – everything takes on a practical and commercial (for money) aspect.
- Americanization through McDonald's, KFC and Starbucks.
- Homogenization – all cities of the world look the same. Experiencing the diversity of building styles, dishes and cultures was once a main reason to visit cities. Today as one travels around the world one finds that many airports, hotels and cities are more or less the same – this takes the fun out of traveling.
- Free flow of money around the world where it can earn the highest rate of return on investment
- The location of manufacturing shifts to where wages are the lowest at the expense of local people. Souvenir production is an example whereby Australian koala toys are manufactured in China or where coconut souvenirs sold in the Maldives are made in Bali.
- Small scale operations such as Peregrine Adventures, a Melbourne based Tour Company, are bought out by bigger companies (First Choice in the UK) who in turn are purchased by even larger companies (TUI of Germany).

- Most of the economic restructuring (change in ownership) is in the interest of big business not in the interest of the public and of the local people.
- The Hilton Hotel Corporation was recently sold to an investment fund that is primarily interested in a high rate of return on its investment for its shareholders and not in the wellbeing of guests, local people, and employees or in corporate social responsibility.
- Multinational companies such as CNN present consumers with their idea of the "truth" behind the news. They tell us what is important and what is not. They show us places we should want to travel to
- National Geographic brings the world's most exotic travel experiences to our living room and we become "armchair travelers." This raises the question whether we still have to travel to places when we have already 'experienced "them on our flat screen TVs.

Globalization is felt particularly in less developed nations, where tourism is frequently seen as a key means of economic development. Unfortunately, in order to establish the necessary infrastructure for the development of tourism, poor countries may need to borrow billions of dollars.

This can be problematic because they may not be able to repay their debts.

Airports, roads, hotels, restaurants, and theme parks may be constructed, but those who cannot afford to go by plane or stay in those hotels will find little use for them. Outside developers are frequently given tax breaks (concessions) as incentives to build hotels and other tourism infrastructure in developing nations.

Although it deprives the local government and the inhabitants of cash by allowing developers to avoid paying local taxes for the first five or more years, there would frequently be no infrastructure built if such concessions were not made.

International travelers demand hotels, food, and drinks in the same style as those found in other countries.

As a result, developing nations are forced to import upscale building supplies like marble from Italy, gourmet foods like oysters from Australia, and alcoholic beverages like Scotch whiskey from other countries, which means that a large portion of the tourism industry's profits go abroad.

As the money can't be used to help the local population and instead leaks out of the local economy, this is known as leakage.

4.4. DEVELOPING TRENDS IN HOSPITALITY EDUCATION – NEW-FANGLED ENCOUNTERS

The act of fostering a bond between a host and a guest is known as hospitality. As a developing country, tourism trends already boost the hospitality sector. As a result, a number of institutes arrived in the front row to participate in the expanding education sector where quality has been bargained ahead of demand.

The hospitality sector is no longer solely comprised of hotels but has expanded to include restaurants, cruise ships, call centers, BPOs, and customer service. The retail industry, communications, hospitals, banks, and so forth. Those who are fluent in Fluent English and eager to help are eligible for these opportunities (Figure 4.2).

Figure 4.2. Call Center.

Source: Image by Flickr

4.5. GETTING EQUIPPED THE FUTURE DYNAMISM OF HOSPITALITY MANPOWER

Diversification is one of the key elements of the current hospitality industry picture, which is booming with numerous domestic and foreign brands.

Therefore, there is a greater opportunity than before and there is a greater need for skilled labor (Gursoy, Rahman &Swangera, 2012).

Although there is a supply of labor, educators must train individuals with the necessary skills, knowledge, and attitudes to fill the gap in order for the industry to remain viable (Amoah & Baum, 1997). Industry analysts suggest that to formulate in accordance with this criteria:

- The existing curriculum must be moderate from time to time by the industry professionals.
- Additional subjects like-accounting, marketing all need a hand to hand training as well.
- Behavior skills training are one of the most principal parts of the approach to building manpower.
- Mindfulness training for faculties is very important to get up to date as per the industry norms.
- The Gov. and Ministry of tourism must take some initiative to track all these to set the high standards (Michael & Manoharan, 2016).

4.6. PRESENT SCENARIO OF HOSPITALITY EDUCATION SYSTEM

The Federation of Hotel & Restaurant Associations (FHRAI) has made the following discoveries: Hotel and restaurant managers from metro areas discussed the need for development and a greater emphasis on practical domains in the underwhelming hospitality education system.

On the other hand, they also had similar thoughts regarding educational systems, qualified faculty, and infrastructures.

Many tiny colleges and institutes, despite their claims to be the best on the market, have been observed to not be kept up to date to provide a minimum level of infrastructure and other amenities.

Along with that, another crucial factor is the caliber of the student, their degree of thought, and their tendency to prioritize their job aspirations on the lower end of the priority scale.

4.7. FACTORS AFFECTING TOURISM INDUSTRY AND ITS IMPACTS ON GLOBAL ECONOMY OF THE WORLD

Traveling from one location to another for fun and recreation is referred to as tourism. It has a significant impact on a nation's progress. Through tourism, there is an increase in the demand for various commodities, and various cultures from across the world are merged into one.

On the one hand, there has been development, but on the other, there is a global spread of various diseases. The tourism business is actually interconnected with every area of the global economy, having both a good and negative impact on the GDP of the nation.

The creation of numerous jobs through the tourist sector has had a significant positive impact on the global effort to reduce poverty and advance socioeconomic conditions. Both the domestic and international industrial sectors are boosted by tourism. There are numerous elements that draw tourists to a place, propelling the global tourism industry.

Natural beauty and historical sites are among the variables that have a big impact on the growth of the tourism sector. In a similar manner, peace and security are crucial for the growth of this industry. Militancy in Pakistan in 2008 and 2009 recently caused significant harm to tourism. Foreigners never entered Pakistan because of terror because militants slaughtered civilians on a regular basis and caused fear throughout the nation.

In Afghanistan and Sri Lanka, similar panics also occurred. The development of the tourism business is therefore impossible without total peace. Every nation in the world strives to grow its tourist industry in order to boost its economy. Recently, the corona virus has posed serious issues for the growth of the tourism industry.

Due to the spread of the Corona virus, every country in the world has declared a state of emergency, and travel outside is prohibited. They have fortified their borders and ordered a lockdown at their homes, preventing anybody from leaving. As a result, the tourism industry has severely harmed the world economy at the moment.

Travelers bring new ideas to other nations, which are vital to the development of a nation, broadening the people's worldview. Generally speaking, tourism helps rural residents' socioeconomic conditions. Tourists visit a place for the natural scenery while also spending a lot of money on various locally produced goods. As a result, cottage industries have grown and the local population has increased income from tourism. Therefore, while tourism on the one hand helps to better the socioeconomic situation throughout the world, on the other hand, their privacy has been compromised, which has spread to other nations. Similar to how one country illness went from one country to the entire world, the corona virus spread from Wuhan, China, to all other countries, causing the global economy to crumble.

No country has been immune to the coronavirus. China has a significant economy and is related to international trade. All sectors are currently closed due to this infection, making it impossible to solve the issue. Similar to how the price of oil has decreased globally, new possibilities have also been opened up.

Therefore, despite occasionally spreading panic and causing the global economy to collapse, tourism benefits the global community. Due to the significance of this topic, the current study was designed to look at the variables influencing the global tourist industry and determine how they affect the world economy.

4.8. TOURISM TRAINING AND EDUCATION IN LESS-DEVELOPED COUNTRIES

4.8.1. Human Resource Development

Governments in many LDCs throughout the world are developing HRD programs as a significant undertaking to attain the more general goals of equitable growth and development (Agrawal, 2012; Hampton & Jeyacheya, 2013; International Labor Organization, 2013). In many LDC today, tourism is a key service industry that offers employment possibilities to women, young people, and people from both rural and urban areas.

It generates about one out of every twelve jobs globally. (United Nations World Tourism Organization, 2010). Tourism is a labor-intensive industry where people from different trades work. A qualified and motivated workforce is essential because the success of the industry as a whole depends greatly on the quality of the services provided (United Nations World Tourism Organization, 2010).

Tourism contributes significantly to economic prosperity in the majority of developing countries, and its growth raises foreign exchange revenues. (Hampton & Jeyacheya, 2013; Knowles, Diamantis, & El-Mourhabi, 2004; Shareef, Hoti, & McAleer, 2008). Intense competition amongst tourist destinations is brought on by the growth of the industry, therefore if destinations want to maintain their market appeal they must be innovative in the items they offer (Hampton & Jeyacheya, 2013).

The tourist sector is dynamic by nature, with popular places coming in and out of style (Hampton &Jeyacheya, 2013This has consequences for the planning and administration of tourism development because product innovations are essential for tourism to be included in inclusive growth programs.

This calls for qualified specialists in the fields of destination marketing, tourism planning, and policy development in order for the development of the tourism industry to be successfully incorporated into national development goals.

Human capital is becoming a crucial element for competitiveness as developing nations increase their involvement in the global economy. For the tourism and hospitality sectors to succeed, it is crucial to improve the skills of a productive workforce. The improvement of skill development for the workforce is greatly aided by training.

Businesses and the individual who work for them must regularly update and enhance their abilities in a quick and efficient manner if they want to stay in the competitive tourism sector market. (Fernandez-Stark, Bamber, &Gereffi, 2012).

In most developing nations, the majority of the workforce is low skilled or untrained, and many of them possess talents that are in low demand. (Agrawal, 2012; Baum, 2006). According to a 2007 study by Akama and Kieti in Kenya, the majority of locals employed in the tourism sector in Mombasa and other Kenyan cities typically hold unskilled, low-paying jobs like waiters, gardeners, porters, janitors, and security guards, while foreign nationals typically hold the lucrative jobs at supervisory and managerial levels.

This study demonstrates the necessity for tourism education and training to increase local capacity and improve the knowledge and abilities of locals working in the tourism and hospitality industries. According to Kaplan (2004), skills development is essential to a successful tourism-led development strategy and can help the national tourism industry change and grow.

In order to ensure the efficient and long-lasting transformation and development of the tourism industry in developing nations, skills development is crucial. The concept of low skills and its applicability to LDC were discussed by Baum (2008), who claimed that the skills that are described as low-skilled in a Western developed context cannot be described in this same way in other countries due to the different experiences in culture, communications, language structure, and relationship assumptions that influence the nature and status of people's jobs in LDC.

Similar research conducted in other LDC environments lend weight to the idea of cultural impacts (Baum, 2006; Lee-Ross, 2001). As an illustration, Lee-Ross (2001) found cultural factors that change between two distinct cultural settings that influence employees' views of their job qualities and internal work motives.

Even international tourism and hospitality companies in LDC will modify Western norms and procedures to fit various local circumstances (H. Adler & Rigg, 2012). LDC training is crucial since the sector changes as new trends emerge and outmoded ones go.

Service personnel need to be aware of the changes and higher standards of customer service (Onyango, Ayieko, Odhuno, &Okech, 2009). Improving productivity requires HR procedures that place a priority on training and development, ongoing learning, and investment in skill development (Hai-yan & Baum, 2006).

This strategy is thought to be crucial for boosting productivity, cutting long-term expenses, and spurring innovation—all of which are advantageous for the tourism industry. Initiatives for skill development through training should strive to provide workers with the necessary skills to meet employment requirements across diverse sectors of the national economy. (United Nations World Tourism Organization, 2010).

For better delivery of tourism and hospitality services in LDCs, training and education in the industry are crucial. The LDC government is redoubling its efforts to support technical and vocational education and training. (TVET) (Agrawal, 2012; Dasmani, 2011). The development of skills is thought to increase productivity and maintain competitiveness in the global economy.

However, despite the fact that qualified workers are in high demand in the majority of countries, companies rarely find applicants with the necessary abilities. The tourism business in LDCs faces a number of serious problems, one of which is that many workers leave the sector in search of better pay and working conditions, taking their expertise with them. (Baum, 2006). According to Wang's (2011) research, one of the challenges facing China's hotel sector is finding workers who are properly qualified to support the industry's rapid growth. Furthermore, Baum (2006) outlined the difficulties that vocational education institutions in LDC face, including a lack of quality, challenges in teaching the skills required by the private sector, and a lack of prestige in comparison to universities because technical education is thought to have low status and low income potential.

In addition, many TVET institutions' tourism and hospitality training curricula need to be evaluated because there is frequently a mismatch between the training provided by these programs and the skills that employers demand (Agrawal, 2012; Mayaka & Akama, 2007).

When managing a big number of unskilled or low-skilled personnel in the industry, finding the correct type of training that matches the demands of business and trainees can be difficult (Teng, 2008). Creating collaboration amongst the many stakeholders can support achieving greater results from practice-based training requirements (Agrawal, 2012).

Unfortunately, industry and trainers do not always work together and cooperate to the same extent in developing nations as they do in the developed world. Hotel managers were shown to be extremely hesitant to participate in external training in Wang's (2011) study conducted in China.

The managers had doubts because they thought that training was an expensive endeavor for their company and because they were unsure of how well the external training and development initiatives would fit into the actual operational environment of their hotels.

Despite an improvement in the quality of formal education in LDC, the tourism industry is still characterized by a workforce with relatively low levels of skills when compared to other economic sectors, and yet there are growing demands for high-quality skills as well as for skills like ICT, which are derived from the developing new job profile (Hai-Yan & Baum, 2006).

Training in tourism and hospitality is hampered and challenged by a number of economic, political, and social variables. These include issues with funding training facilities, a lack of understanding of tourism in the secondary school system, and problems recruiting qualified teachers (Imbal, 2005; Zagonari, 2009).

Additionally, there are outside variables that influence a student's likelihood of finishing their education and training, such as financial demands and other outside variables (Huang & Busby, 2007). Another element associated with students not finishing their courses is individual learning styles. According to Teng's (2008) research, a student's personality is a good predictor of their attitudes regarding career aspirations in the tourism and hospitality sector.

Last but not least, incorporating experiential learning into higher education curricula can help students develop their practical abilities and, in turn, increase their chances of finding work (Yan & Cheung, 2012).

4.9. EMERGING TRENDS IN HOSPITALITY EDUCATION: NEW CHALLENGES

The term "hospitality" refers to both the art and practice of being hospitable as well as the relationship between a host and a guest. Although the country's economic liberalization, open skies policies in the aviation industry in the early 1990s, and later effects of globalization gave the industry a new boost, the proliferation of hospitality institutes has posed some risk to the quality of education as many institutes are being run without any proper infrastructure.

For any business to succeed today, providing excellent customer service and ensuring that customers are satisfied. When they graduate from hospitality schools, students learn how important they are in a variety of non-hospitality sectors, including retail establishments, healthcare facilities, and private banks, the telecom sector, food packaging facilities, cruise lines, and BPOs.

Many hospitality management colleges have formed strategic alliances with numerous foreign universities to make student transfers, faculty and student exchange programs, curriculum collaboration, and other initiatives possible.

Universities and institutions where those students from various institutes can finish their internships. Unfortunately, there are organizations that collaborate with lesser-known universities solely for the final year of study abroad, demanding fees in the amount of Rs 10,00,000 in the name of an internship.

It is time to sincerely consider whether sending students abroad for internships benefits them in any way, or whether we are only providing cheap labor under the guise of an internship.

4.9.1. Challenges of Hospitality Education

- The curriculum and course materials determine whether a course is successful. Since the needs of the hospitality sector are changing quickly, the curriculum must reflect these changes. The syllabus committee must include people from the industry (ideally general managers or operation managers) to assist in creating the syllabus.

- The required steps have been made by AICTE, universities, and NCHM to restructure their curricula in order to provide a value-based curriculum. Unfortunately, despite this, hospitality syllabuses are not uniform throughout the country.
- In certain states, the State Council diploma course syllabus is substantially more detailed than the degree course syllabus and includes several peripheral and non-core subjects. The importance of task-based and management-based subjects must be emphasized. Personal growth, leadership development, grooming, and other issues of additional value should be covered in the curriculum.
- The curriculum should include fresh subjects like call center management, retail store management, and hospital management. In order to provide a unified curriculum, proper coordination between the organizations that approve courses, such as NCHMCT, AICTE, institutions, and others, is required. Organizing all hospitality courses under one roof will allow for sufficient oversight and uniformity.
- Technology has ingrained itself deeply into our daily lives. It has become essential to manage hotels using technology. Knowledge of these technologies particularly in property management is very important. The requirement of the hour for creating presentations, teaching, conducting research, etc. is to place more emphasis on the use of computers and PMS software at the institute level. In the kitchen, restaurants, cleaning, and front office divisions, curriculum should promote increased usage of and familiarity with contemporary equipment.

4.10. ACCOMMODATION SECTOR

4.10.1. Disposable Income

Economic instability has a significant impact on the hospitality sector since consumers depend on their disposable income to cover their trip expenses. RevPAR (Revenue Per Available Room) and GDP have historically had a similar economic relationship (HotelNewsNow, August 2014).

According to Deloitte's report on Hospitality 2015, the overall decline in hotel demand during the recession of 2009 was the most pronounced decline ever recorded over the past century.

4.10.2. Implications for Types of Accommodation

As a result of the worldwide economic slump that began in late 2008, the lodging sector was significantly impacted. The economic downturn led both business and leisure tourists to reduce their travel expenses.

They have developed a greater awareness of the expense of travel, and as a result, they have sought out more affordable options. They have also stayed at low-cost hotels and used low-cost transportation services. Following the recession, people are seeking greater value when making purchases.

4.10.3. What Are Select-Service Hotels?

According to the Manual of Introduction of Hospitality on Hotels and their classifications, there are nine types of classification factor:

- Market Segment
- Design
- Function
- Staff to room ratio
- Rating
- Location
- Distinctiveness of properties
- Size
- Price

Based on the above classification factors, Select-service hotels are classified by the price factor. In general, hotels under this classification (price) factor can actually further break down into three main classes, and Select-service is one of the three classes (Figure 4.3):

1. Limited-Service
2. Select-Service
3. Full-Service

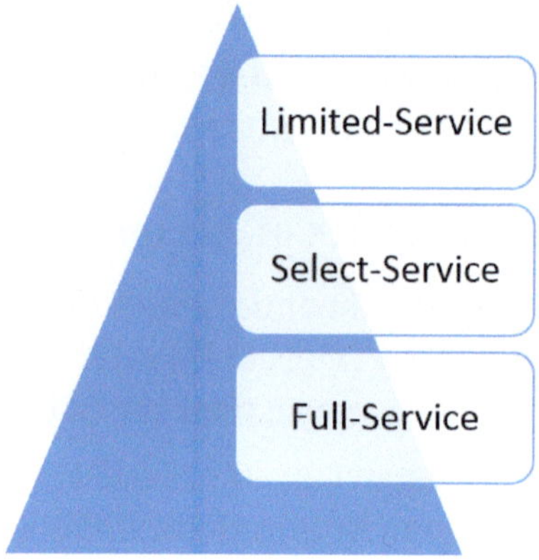

Figure 4.3. Select-service hotels.

a. **Limited-Service Hotel**

Initially, a hotel without a restaurant or a banquet space was considered a limited-service hotel. In general, limited-service hotels offer straightforward amenities and services to their guests. It should be emphasized that these offerings have grown over the past 10 years.

A limited-service hotel may now include amenities including a business center, a fitness center, a guest laundry facility, a market pantry, an indoor and/or outdoor pool, and small conference rooms. Limited-service hotels don't provide catering from various restaurants.

Due to the demand for limited-service hotels typically coming from budget-conscious business and leisure travelers, room rates are also typically on the lower end of the spectrum.

b. **Select-Service Hotel**

A hotel that offers both limited and full services is known as a select-service hotel. It provides the essentials of limited-service hotels as well as some of the amenities and services that are features of full-service hotels.

This typically means that these hotels lack the extensive catering options, numerous restaurants, and meeting space that one would typically find in full-service hotels. However, while keeping costs down, in-room amenities can come close to or even match those offered at full-service hotels.

In fact, select-service hotels have experienced an increase in commercial demand as a result of tighter corporate travel budgets both during and after the 2008 global economic downturn. Overall, it is reasonable to conclude that select-service hotels are more comparable to limited-service hotels, but individual select-service hotel offerings can differ.

c. **Full-Service Hotel**

The abundance of food and beverage options that are acceptable for both guests and groups is the most distinctive aspect of a full-service hotel. Further, certain features like spas, expensive event halls, doormen, valet parking, extended room services, concierge services, high-end restaurants, and boutiques set many full-service hotels apart from one another. The extra amenities and higher level of service only available in full-service hotels are generally sought after by these travelers (Figure 4.4).

Figure 4.4. Full-Service Hotel.

Source: Image by Needpix

4.11. CONCLUSION

In this chapter issues and challenges in tourism and hospitality education. Tourism industry play great role in the development of a world and it positively and negatively correlated with the economic growth of the global economy. Hospitality education has appeared as an influential power for increasing economies. This has been accepted as a medium for visualized development and growth. On one side development occurred but on the other side different diseases are transferred from one country to another in the world.

Towards the end of this chapter, this chapter also explains the tourism training and education in less developed countries among many of the world's LDC, governments are creating HRD plans as a major initiative to achieve the broader objectives of inclusive growth and development.

REFERENCES

1. Bauer, D. T., (2014). *Trends and Issues in the Tourism and Hospitality Industry*. https://www.edb.gov.hk/. The Government of the Hong Kong Special Administrative Region. Available at: https://www.edb.gov.hk/attachment/en/curriculum-development/kla/pshe/references-and-resources/tourism/Final_Trend_English_a3.pdf (accessed on 14 June 2023).
2. Khan, N., et al., (2020). *Factors Affecting Tourism Industry and its Impacts on Global Economy of the World*. https://papers.ssrn.com. SSRN. Available at: https://deliverypdf.ssrn.com/delivery.php?ID=13811202409412601 21090140220280970930030820270010190580670980090180040990860100940980431030400380560110 07 069028003127030089004023055074034000001087120115031 08611608707103903511311212609902507012507909601601901311302 5030 077004114009087122076103118031087&EXT=pdf&INDEX=TRUE (accessed on 14 June 2023).
3. King, B., (2015). *Tourism and Hospitality Education*. Tourism Management. Available at: https://www.academia.edu/11274933/Tourism_and_hospitality_education (accessed on 14 June 2023).
4. Kumar, M., (2014). *Hospitality Education in India-Present status, Challenges and Opportunities*. ajhtl.com. Maharshi Dayanand University. Available at: https://www.ajhtl.com/uploads/7/1/6/3/7163688/article_10_vol._3_2_july_14.pdf (accessed on 14 June 2023).
5. Mayaka, M., (2017). *Challenges for the Tourism, Hospitality and Events Higher Education Curricula in Sub-Saharan Africa the Case of Kenya*. Academia.edu. Academia. Available at: https://www.academia.edu/30866420/Challenges_for_the_tourism_hospitality_and_events_higher_education_curricula_in_Sub_Saharan_Africa_The_case_of_Kenya (accessed on 14 June 2023).
6. Murki, N. Y. S., (2014). *Challenges of Training Tourism and Hospitality Workers in Papua*. https://openrepository.aut.ac.nz. Openrepository. Available at: https://openrepository.aut.ac.nz/bitstream/handle/10292/8334/MurkiNYS.pdf;sequence=3 (accessed on 14 June 2023).
7. Roy, A., (2021). *Issues and Challenges of Hospitality Education System in India*. https://www.researchgate.net. Bharti Publications. Available at: https://www.researchgate.net/publication/353547115_Issues_and_

Challenges_of_Hospitality_Education_System_in_India (accessed on 14 June 2023).

8. Skripak, S. J., (2016). *Hospitality and Tourism*. vtechworks.lib.vt.edu. Pamplin College of Business and Virginia Tech Libraries. Available at: https://vtechworks.lib.vt.edu/bitstream/handle/10919/70961/Chapter%2015%20Hospitality%20and%20Tourism.pdf (accessed on 14 June 2023).
9. Thakur, S., & Kulshrestha, D. R., (2020). *Tourism and Hospitality Perspectives, Patterns and Practices*. ResearchGate. Bharti Publications. Available at: https://www.researchgate.net/profile/Abhishek-Roy-20/publication/353547115_Issues_and_Challenges_of_Hospitality_Education_System_in_India/links/611b38f01ca20f6f862981e2/Issues-and-Challenges-of-Hospitality-Education-System-in-India.pdf (accessed on 14 June 2023).

Chapter

5

EMPLOYABILITY AND SKILLS IN TOURISM AND HOSPITALITY

Contents

5.1. Introduction to Employability and Skills 110
5.2. Vocational Employability Skills 111
5.3. Graduate Employment 113
5.4. Types of Employment: SME, MNC, Non-Profit, Public, and Self 115
5.5. Employment in Tourism and Hospitality 117
5.6. Employment Opportunities in Tourism and Hospitality Sectors 119
5.7. Essential Skills for a Career in Tourism and Hospitality 122
5.8. Employment in the Tourism Industry to Grow Significantly 128
5.9. Challenges of Employments in Tourism and Hospitality 131
5.10. Future of Employment in the Hospitality Industry 133
5.11. Conclusion 137
References 138

In the chapter of employability and skills in tourism and hospitability talks about the vocational employability skills and graduate employment. This chapter also explains the different types of employment such as – SME, MNC, Non-profit, Public and Self. This chapter provide highlights on the employment opportunities in tourism and hospitality sectors. The chapter also explains about the essential skills needed for a career in tourism and hospitality.

The chapter also mentions the challenges of employment in tourism and hospitality sectors. At the end of the chapter, the chapter explains about the future of employment in the tourism and hospitality industry.

5.1. INTRODUCTION TO EMPLOYABILITY AND SKILLS

Employability Skills are transferrable abilities an individual must possess in order to be considered "employable." Employers frequently list a specific set of talents they are looking for in a candidate, in addition to outstanding technical comprehension and topic knowledge.

They feel that having these abilities will enable the employee to perform their job to the best of their ability. Your knowledge, skills, and attitudes, as well as how you use those assets and how you portray them to employers, all affect your employability.

Your education and experience may qualify you to apply for a job, but in order to succeed in most positions, you will need skills that you will probably acquire over time. While some will be job-specific, the vast majority will be what are known as "soft skills," which can be applied to any position or field of employment. These "employability skills" are what make you employable

Employers are typically eager to teach employees the job-specific skills necessary, such as how to operate certain machinery or utilize specific computer programs that are extremely specific to that role or business. The other "soft skills" are significantly more difficult to teach, therefore employers typically want to see that you already possess them before hiring.

Employability skills include things like:

- good communication
- motivation and initiative
- leadership

- reliability/dependability
- following instructions
- team work
- patience
- adaptability
- emotional control
- resilience

Employable:

1. Suitable for paid work.
2. Able to be used.

Skill:

1. The ability to do something well: expertise.
2. A particular ability.

The aforementioned dictionary definitions provide helpful starting points for thinking about what it means to be employable and skilled. However, it must be noted that employing dictionary definitions in academic work is regarded as poor quality. Prior to conducting academic study, this acknowledgement of these is provided to clarify the standpoint from which people may typically comprehend the words.

In order to define employability and skills for jobs in the tourism, hospitality, and events management industries, it is crucial to make it clear that these positions call for specialized education and training, as opposed to the conventional training that is typically provided for non-vocational positions.

Having said that, there are positions in the travel, hospitality, and event industries that could be viewed as non-vocational.

5.2. VOCATIONAL EMPLOYABILITY SKILLS

The "social development of labor" is achieved through vocational education (Clarke and Winch, 2007: p.1). It is referred to as vocational education and training (VET), and its main goal is to get people ready for the workforce. This is different from liberal or civic education in that it calls for students to learn particular competences and skills for a particular job.

Vocational skills are frequently mentioned as being necessary for successful employment. Because these abilities must be applied practically during training, Moodie (2002) emphasizes that there are difficulties in the connections between traditional higher education knowledge and skills and vocational employability skills.

Comparing vocational training to regular academic education, there is a perception that VET is less scholarly, which is viewed as snobbish. This chapter will explain how this textbook will assist employability (vocational) skills by implementing academic frameworks, which support the critical thinking and academic knowledge required in employment and education.

In various debates of vocational education (*see* Ewert, 1991; James, 1995), Habermas' theory on ways of knowing has been cited because it clarifies the connection between information learned and then applied in the actual application of skills for employment.

The evaluation of skills and the identification of acceptable tasks to perform while keeping information in mind might help identify technical reflection. For instance, an employee is knowledgeable about possible responses to a client complaint (Figure 5.1).

Figure 5.1. Women in vocational education.

Source: Image by Flickr

The employee can then select the best course of action for the particular situation by engaging in practical reflection. Therefore, in the case of a customer complaint, it is important to understand the situation's circumstances and context in order to resolve it within the current corporate, societal, and social contexts. Therefore, critical reflection is the higher level of ability to think critically about many perspectives and results from a scenario (individually and within team or organizational contexts).

This acknowledges the conflict between applying academic knowledge to real-world situations and utilizing talents. Additionally, it is acknowledged that traditional educational approaches can frequently result in the development of occupational skills (see a research case study from Malaysia by Bakar and Hanafi, 2007).

It is evident that VET enhances employability skills and can also enable a person to critically evaluate their knowledge of skills in employment contexts by recognizing the nature of vocational education and locating it within Habermas's knowledge definitions. utilizing a variety of academic frameworks (knowledge), and by challenging readers to critically reflect on them through activities, ponder points, and revision questions.

5.3. GRADUATE EMPLOYMENT

The ability to work or travel while being compensated or trained for a particular industry or job is known as graduate employment. It indicates your capacity to secure a contract of employment after completing your education, but it has nothing to do with a specific pay or degree of employment.

After finishing a course, many of my students say they feel pressured to land full-time, well-paying management jobs. As a lecturer at a UK university, it's important for me to remind my students that their transition into adulthood will be unique, based on their personal goals and the economic chances and labor market where they choose to find employment. Employers will anticipate you to possess a specific set of abilities, competences, knowledge, and personality attributes as a "graduate.

"It's critical to comprehend what you as a graduate are actually supplying if you want to comprehend what companies think of as a graduate characteristic. Within the UK there is the Quality Assurance Agency (QAA), in Australia there is the Tertiary Education Quality and Standards Agency (TEQSA) and in America there is the Council for the Advancement of Standards in Higher Education (CAS).

These particular benchmarks are present in every nation that offers higher education to make it clear what level of competency you will reach when you meet the requirements.

If you were to conduct a search for "graduate employability abilities," you would come across a vast amount of academic research and conclusions that attempt to connect graduate traits to employer demands across a wide range of businesses and sectors.

There is a huge variation in the recommendations from research in THE. Some claim that the most crucial abilities are those of leadership and interpersonal communication (Kay and Russette, 2000), while others assert that there is no correlation between graduate and industry capabilities (Jackson, 2012) or that there is a clear parity between management and student abilities (Suh et al., 2012).

According to UK prospects from 2009, all graduates should have self-reliance, interpersonal, general employment, and specialized skills. Who is correct, then? What skills do you need to develop in these areas? Help!

Please be aware that there isn't a one, precise formula for what THE companies want from you as a graduate. Every business seeks employees with the abilities necessary to do tasks in the most productive and affordable way, as was stated at the beginning of this section (Figure 5.2).

Figure 5.2. Interpersonal skills between two person.

Source: Image by Flickr

It is crucial to describe different graduate pathways after finishing your studies before moving on to the types of businesses in which to seek job for THE.

This textbook will provide you the opportunity to evaluate your abilities and comprehend important concerns in work within THE, but this is viewed as an ongoing and lifelong process. You can decide that after graduation you wish to travel or pursue more education. Successful graduate employment outcomes can be achieved in each of these directions.

5.4. TYPES OF EMPLOYMENT: SME, MNC, NON-PROFIT, PUBLIC, AND SELF

Knowing what kinds of companies offer THE jobs and determining whether these companies are suited for you is crucial for developing your employability. We'll first look at the background and legal definitions of employment. Following that, the definitions of the THE sectors' terminology for small- to medium-sized businesses (SME), multinational companies (MNC), non-profit, public, and self-employment will be given.

Over the years, working procedures have evolved. Roman era saw the beginning of payment for labor. The word "salary" comes from the Latin word "salarium." However, throughout the Roman era, there were no pounds, dollars, or euros.

As an alternative, people would barter or work in return for goods. We didn't begin using currency until economists like Benjamin Franklin and Adam Smith began to question the theory of labor and the values attached.

These two quotes from Franklin's article show that paper money first appeared and was used to pay for goods and services in the 1700s. The first quotation demonstrates that money was being used in place of gold "bullion." Money was therefore valued according to the amount of labor necessary to produce an equivalent amount of metal or land.

Secondly, the value of money as a medium of trade increased in proportion to the length of time it took to conclude the negotiations over the terms of the work's payment. Today, money is utilized as payment for services, but there are other distinctions such as minimum wage, tax, insurance, money, and contract type.

It is crucial to take into account the different types of employment contracts that exist after noting the history of payments made in exchange for labor. This is crucial for THE industries since managers depend on a diversity of staff members. First of all, every country has different job contracts. People who work in the UK are either called employees or workers (Government, 2012). A person who has a full-time or part-time contract who consistently works for the same business until the employment contract is canceled is referred to as an employee in the UK.

A worker's employment contract also has a start and end date. Employees are also sometimes referred to as temporary staff. Casual employees are a third class of employees in the UK. Although they work for just one company, these individuals may not have a set number of hours to work. These agreements are sometimes referred to as zero-hour agreements.

In America, at-will employment agreements are common (Feinman, 1976). Here, the job relationship can be ended at any time without consequences for either the employer or the employee. A contributing element to flexible contracts for THE, according to Baum (2006a), will be the demand for flexible working hours for the employees and from the staff, even though different nations will offer a variety of employment contracts.

Now that you have a general understanding of how to get paid for your labor as well as the various employment arrangements and contracts, it is crucial to comprehend how they vary between THE SME and MNCs.

Any company with less than 250 people and an annual revenue of less than €50 million is considered a small- to medium-sized enterprise (SME) (Enterprise and Industry, 2012). Due to the assistance offered to these smaller businesses, this sort of business is designated by EU law. There are additional risks for the directors to address because a SME is a small-sized business.

As they assist the MNCs, they are a crucial component of any economy. The corporate giants of business are multinational corporations (MNCs). These can be found in great numbers if you go down any high street in any significant city.

Working for a multinational corporation will give you training and experience from a prestigious organization that future employers will value. Typically, these positions are full-time and long-term (no fixed end date).

However, they can have stringent requirements and procedures for finishing your work. This can be helpful when learning about a job in the industry, but it may not be appropriate for all employees. For instance, if you want to start your own firm, it will be beneficial to work for a comparable MNC to learn how a successful organization runs before starting your own SME with the knowledge learned. Beware though, some employment contracts forbid you from using their intellectual property or stealing their clients after your work has ended Even while you might be a bigger fish in a SME's pond, this will also mean you have more responsibility and are more noticeable.

Seasonal employment with a set end date is frequently preferred in THE SME businesses because it allows them to better control the cost of hiring new personnel. This also means that face-to-face contact and who you know is more important.

Additionally, THE graduate employment can be carried out as self-employment. This is customary in the field of event management, where a number of employees carry out short-term contracts for various festivals and events. But in this chapter, I'm going to explain how you could opt to launch your own firm and provide help for THE companies.

You will have critically examined important jobs and processes within THE employment practices during your time studying THE. You might have discovered a personal interest in this in the administration of specific leisure activities.

With this interest and knowledge, you are in a unique position to start and build your own business. After graduation, you should consider taking on this risk-intensive profession because you will be in charge of managing your own time. However, it will allow you to make a great salary and feel quite satisfied with your job.

5.5. EMPLOYMENT IN TOURISM AND HOSPITALITY

5.5.1. Role

The role and level you choose to enter will depend on your own competences, competency and comfort zone. Traditionally in THE, there are four levels:

1. Floor/office staff.
2. Supervisors.
3. Managers.
4. Area managers.

Each of these roles can be found in the company's various departments. For example, at a music festival you will see bar staff

1. supervisors of the bar
2. bar management on site
3. and then brewery managers who supply the alcohol
4. I frequently hear from students or recent graduates that they don't feel confident applying for management or area management posts until they have graduated or worked a number of years in full-time employment. What do you have to lose?

Is always my response to this question and concern? If you wish to work in a specialized field of tourism, like managing a heritage site in Malta, you can be very interested in the position but not quite confident in your managerial abilities. Prior to completely dismissing a position, I would always suggest calling the employer for a casual discussion about the position.

5.5.2. What Suits You Best?

It is important to think about the roles that currently suit you before analyzing your existing skill set. Some companies will ask you to complete a psychometric test so they can evaluate this. An employer can use psychometric testing to determine whether your existing skills, abilities, or personality match those desired by the organization at the time. There are several tests readily available online for you to complete, and you may take them whenever you like.

Cubiks (http://practicetests.cubiks.com/) offer a numerical and verbal test for business supporters, managers and graduates.

Keirsey (www.keirsey.com) tests can be completed to show you your own temperament for certain roles. However, these tests will only provide a limited understanding of the type of employee you are at the moment. (I am suspicious and reticent to say use them absolutely and base your own career direction on them).

Following some lighthearted experimentation and self-discovery with these online tests, I would advise giving some thought to the following questions:

1. Do you live to work or work to live?
2. Do you want to travel?
3. Do you want to develop or maintain your skills?

Answers to these questions will enable you to define your current suitable role. If you work for the sake of working, a position with many hours, loads of responsibility, and pressure will suit you. If you want to travel, think about working for an MNC because you'll get employee perks and travel chances. Look for work that is comparable to the part-time work you did while attending college or university if you want to keep using your skills. To put it simply, "employable" and "employability" refer to whether or not a person has the necessary abilities to find or keep a job. You may already have a part-time job as a student, which makes it simple for you to find temporary, flexible, or casual employment. However, it's crucial to demonstrate experience, talent, and expertise to land a job if you graduate or start looking for more long-term, managerial responsibilities.

5.6. EMPLOYMENT OPPORTUNITIES IN TOURISM AND HOSPITALITY SECTORS

One of the biggest and fastest-growing industries worldwide is tourism. The three key industries that drive the global economy in the twenty-first century are technology, telecommunication, and tourism. Tourist attractions, transportation, lodging, dining and drinking places, retail stores, entertainment, business, and other hospitality services offered to people traveling alone or in groups away from home are all included in the category of tourism. The total economic impact of travel and tourism increased to 9.5% of worldwide GDP in 2013. Travel and tourism supported about 266 million jobs in total in 2013 (8.9% of all employment),(1 in 11 of all jobs in the world) .

Figure 5.3. Panel discussion on the organization of work and production, The Future of Work We Want: A Global Dialog.

Source: Image by Flickr

Despite being impacted by the global economic and social crisis, international tourism is expected to generate 296 million jobs by 2019 according to the ILO Global Dialog Forum on New Developments and Challenge report. As a result, tourism has the potential to be a significant job creator (Figure 5.3).

As a major employer and source of labor, tourism is very labor-intensive. It ranks among the top employers in the world for occupations requiring a range of skill sets and enables migrant workers, women, and young people to enter the workforce quickly.

In some nations, the contribution of tourism to job creation is important; for example, India is the second-largest employer in the tourism sector, creating up to 25 million jobs. In Dayananda, Chandrakanta Sahoo is mentioned. In Ethiopia, travel and tourism contributed ETB 91, 898.4 million (9.3% of GDP) in total and 2,291,500 jobs (8.5% of total employment) in 2014.

The tourism and hospitality sectors generate numerous job possibilities across a variety of sectors, including lodging, transportation, and attraction locations.

Therefore, the availability of qualified and trained labor is a critical component in the success of any tourism development strategy or program, making employees a necessary component of the tourism business.

However, employment in the tourism sector is restricted by unpredictable employment, low job status, lengthy, solitary working hours, and low compensation. The most visible and immediate effects of such a situation are the difficulty in finding qualified employees and the high staff turnover rates, both of which are costly to the industry's development.

As a matter of fact, workers in the tourist and hospitality industries have received less than adequate attention. However, a great and efficient service delivery system is built around the staff. These employees are therefore possibly the most important resources for the company.

There are two basic ways in which the connection between human resources and tourism might be characterized. First, the tourist sector can only grow if it can employ a sufficient number of high-caliber workers. The third concern is that how employees are handled by managers and customers, and how they are treated by employees, should be consistent with social equity and fairness. Professionalism among personnel has a significant impact on how well tourists perceive a destination and their experiences there.

Employment possibilities in the tourism and hospitality sectors can be developed directly or indirectly depending on the engagement or contribution of the tourism supply side. All of the job opportunities supported directly by travel and tourism are referred to as direct employment opportunities.

For instance, working for hotels, restaurants, travel agencies, tourism offices, museums, protected areas like national parks, palaces, religious sites, and monuments, as well as for airplanes, cruise ships, resorts, retail stores, souvenir shops, photography studios, sightseeing tours, and farms, bed and breakfast inns, and guesthouses, regional transit (state owned airlines and railways, private transport facilities), Scouts, chefs, and guides.

Indirect employment is also supported by tourism and hospitality through businesses that build and maintain tourist facilities as well as the essential infrastructure, manufacture aircraft, create a variety of handicrafts, work with marketing firms, and provide accounting services. These businesses are more or less reliant on the businesses that provide direct employment for their income.

Tourism's contribution to the economy is assessed by how it affects things like income, employment, investment, and development, as well as the country's overall budget.

In a sector with high labor costs, like tourism and hospitality, a larger amount of income is likely to come from wages and salaries paid to people working in positions that either directly serve tourist demands or indirectly profit from their spending.

If there are numerous possibilities and activities for visitors to engage in, income will be high in tourist areas that draw big numbers of people; where visitor length of stay is maximal, customer spending is very high.

The popularity of tourism and international travel around the world, the fact that people now view travel as a human right, the concern that destinations have for providing high-quality services to visitors, the fact that vacationers have more options than ever before, the development of international standards and quality assurance systems by national and international tourism organizations, and a variety of other factors are all contributing to the diversification of employment opportunities.

The cost of travel has decreased as a result of the ongoing reduction in the price of oil internationally. The main causes for the increasing tourist flow in the destinations are all of the aforementioned situations. As a result, a large staff is needed to supply services to tourists. Because of these realities, the tourist and hospitality sectors, along with its enabling industries, offer a variety of employment opportunities.

5.7. ESSENTIAL SKILLS FOR A CAREER IN TOURISM AND HOSPITALITY

Are you an effective time manager who can remain calm under pressure and solve problems with empathy? If so, you should probably consider careers in hospitality and tourism. As more hospitality and tourism companies acknowledge the value of "soft talents," your skills may become highly sought-after in the industry in the years to come.

Frequently, recruiters in the hospitality and tourist industries will spend money on training a less-experienced applicant if they believe she or he has the necessary personality and soft skills for the position.

Think you've got what it takes? Hospitality is a significant industry since there are many employment available and it has a good economic influence on local areas. There are hundreds of jobs available, and many people prefer to stay in the industry for a long time and climb the ladder because it is so diverse. You can work in hotels, catering, beverages, cruises, events, or the nightlife, among other industries.

However, there are a number of fundamental talents you'll need to have if you want to succeed. After all, hospitality is primarily about giving customers exceptional service and making them smile, which is a role that isn't necessarily suitable for everyone (Figure 5.4).

Figure 5.4. Working in cruise.

Source: Image by Flickr

5.7.1. Empathy And Emotional Intelligence

The future of the hospitality and tourist sector will undoubtedly be shaped by digital revolution, but that does not imply it will become impersonal. Contrarily, making an investment in human capital can help you come up with original solutions for problems that are constantly evolving. People determine an organization's performance, and if this is true in all disciplines, it is crucial for service-based businesses like hospitality and tourism.

5.7.2. Teamwork

The capacity to fit in with the existing team is frequently regarded as crucial when choosing a candidate, among the specific qualifications required by a wide range of host enterprises and organizations. The internship and placement department at IULM provides counseling services in order to meet this requirement. These services are targeted at helping students with their interview preparation and making sure they receive the proper mentoring while they are interning. "We help students improve their personal skills," explains Cristina Sottotetti, IULM internship and placement director, "which are […] necessary […] to achieve professional success."

In the hospitality industry, no matter what position you have, you will always be just one part in a much bigger machine. You must be able to collaborate effectively with others, especially during busy times, whether you're working within a specific hotel department, in a busy kitchen, or as a member of the bar staff. This can be difficult given the high rate of industry turnover. After all, you'll have to continually adjust to new people and form relationships from start. However, if you're not operating in unison effectively, everything could start to go wrong, which would be bad for clients, your job, and ultimately you (Figure 5.5).

Figure 5.5. Teamwork Training.

Source: Image by Stockvault.net

5.7.3. Stress and Time Management

Hospitality managers frequently handle multiple tasks at once while handling a large workload quickly. When you are required to complete numerous tasks simultaneously in a short amount of time and are confronted with unforeseen issues, it is simple to let your emotions run wild.

If you want to increase productivity and client satisfaction in employment in hospitality and tourism, you need to be able to multitask and keep your composure.

5.7.4. Problem-Solving

One of the most valued soft skills for careers in hospitality and tourism, and undoubtedly a determining factor when it comes to professional success, is the capacity to think quickly and suggest workable solutions to problems, whether you're dealing with a difficult client or confronted with internal issues.

Regardless of your area of expertise, you must understand how to handle a crisis, from identifying the issue to assessing how well you handled it and what could be done better in the future. This is a skill that is highly respected in every field, but in the hospitality sector, being able to problem-solve fast can help you save a lot of potential headache (Figure 5.6).

Figure 5.6. Two employees sat together working on problem solving skills. One is using a laptop while the other speaks, and they are surrounded by puzzle pieces.

Source: Image by Flickr

For instance, if a customer complains about their accommodation, you may give them a free drink at the bar while you wait for another customer to check out. This avoids a potential problem and keeps the client satisfied while also giving a positive impression of the property. As an alternative, if a client has particularly specific dietary needs, you might speak with the chef on their behalf and provide a specially designed alternate remedy.

5.7.5. Strategy and Innovation

Travel and hospitality Candidates that reject the adage "if it ain't broke, don't fix it" and are willing to go above and beyond innovating are frequently favored by human resources managers. "At Starwood [Hotel and Resorts]," explains Arianna Agresti, Starwood Milan Area HR Supervisor, "we look for talents with strong team-working skills and this is yearly measured through our apposite survey."

She adds, "Among strategic and innovative skills we mention the 'going beyond borders' attitude: the ability to keep looking for new opportunities and solutions even if there are established procedures."

5.7.6. Customer Service Skills

Your ability to fulfill customer expectations is the one thing that, above all else, may make or destroy you in the hospitality industry. Whether you're just pouring beverages or managing a whole hotel, it's your responsibility to make sure that your clients are having a nice time and aren't concerned about anything.

In essence, providing good customer service involves being proactive and positive. Even when dealing with a difficult client, it's necessary to maintain professionalism, smile, and be nice; however, there are times when going above and beyond for a client or guest is also appropriate.

Always keep in mind that you are more likely to receive favorable feedback and a helpful advice if you deliver a great experience.

5.7.7. Interpersonal Skills

Customer service and interpersonal abilities are closely related. You must regularly engage with and help people if you want to pursue a front-facing position in the hospitality industry. Therefore, you need to have interpersonal abilities including empathy, active listening, bargaining, and conflict resolution.

This group of social skills will come in handy when you have to deal with upset or irate customers since they will enable you to defuse situations and handle their problems in a way that will please irate consumers.

5.7.8. Multitasking Skills

The fact that it's so busy practically all the time is one of the reasons why working in hospitality can be so challenging. The ability to multitask and manage multiple activities at once will be useful to you because, in most circumstances, there is no such thing as a quiet day at work.

This requires developing good time management and prioritization skills. You also need to be able to handle stress and maintain composure under duress. These are important soft skills that are highly valued in any employment, even if it's just a part-time job while you're in school (Figure 5.7).

Figure 5.7. Multi-Tasking Efficiency Manager.

Source: Image by Needpix.com

5.7.9. Communication Skills

Every industry values good communicators, but the hospitality and tourist sectors are particularly respected. It's critical that you have the ability to communicate in a way that is both clear and understood and indicative of your employer's brand because you will be interacting with individuals of all ages, temperaments, countries, and backgrounds every day.

As already discussed, the capacity to develop and maintain relationships can significantly impact your ability to attract and retain clients. Additionally, it's critical to be able to speak properly with your coworkers, particularly in hectic, high-stress settings like kitchens or nightclubs where productive teamwork is essential.

5.7.10. Compliance Knowledge

A certain level of education and compliance-related knowledge are required for the majority of hospitality jobs. It's possible that your position will require you to be well-versed on laws that apply to your particular business.

For instance, if you operate in a kitchen, you must be knowledgeable about health and safety laws and make sure that supplies are maintained properly, workstations are carefully cleaned, and food safety protocols are followed.

5.7.11. Foreign Languages

Language proficiency is a big advantage in this sector even though it isn't strictly necessary because it enables you to interact with a larger spectrum of clientele.

If you wish to work in the tourism industry, where the ability to communicate in another language is vital on a daily basis in person, they will be especially helpful.

Long-term career advantages might also come from language abilities. For instance, if you know French, there might be more senior-level operations or management options accessible to you, such as in a customer liaison or relationship management post.

5.7.12. Professionalism

It's critical that you always maintain a high level of professionalism because the majority of companies in the hospitality sector rely on their customer-facing staff to uphold the reputation of their company.

Typically, this entails making sure you are neat and well-groomed, arrive on time for your shifts, and are not observed engaging in any inappropriate behavior (such as smoking outside the main entrance or not washing your hands before handling food). Additionally, it entails maintaining your composure and refraining from retaliating negatively while dealing with a frustrated or furious customer, particularly at the end of a long and tiring shift.

5.7.13. Attention To Detail Skills

Although they won't make or break your job in hospitality, your attention to detail abilities will occasionally come in helpful. The little things can have a significant impact, whether it's catching billing or administrative mistakes at the front desk or realizing that a certain product has passed its best-before date.

Additionally, it might assist you in forging bonds with clients and provide a more satisfying experience all around. For instance, knowing how a certain customer likes their drink to be served, recommending a certain wine to go with a dish, or even observing that someone is having difficulty carrying their luggage and offering to help are all minor details that can have a great impact on customers.

5.8. EMPLOYMENT IN THE TOURISM INDUSTRY TO GROW SIGNIFICANTLY

In the current global economy, one of the biggest and most dynamic industries is travel and tourism. More than 235 million jobs, or 8% of all jobs worldwide, are anticipated to be created in 2010, contributing almost 9% to total GDP.

At the ILO's Global Dialog Forum on New Developments and Concerns in the Hospitality and Tourism Sector in November, more than 150 government, employer, and worker delegations representing more than 50 nations met to explore these developments and challenges.

The International Labor Organization (ILO) Director-General Juan Somavia and UNWTO Secretary-General Taleb Rifai launched the Forum. More than one-third of all global services trade is accounted for by this sector, which is one of the fastest growing when compared to other economic sectors.

The ILO Forum addressed the industry's high labor intensity, which makes it a key source of employment and one of the world's top producers of jobs with a range of skill requirements and easy entry into the workforce for young people, women, and migrant workers (Figure 5.8).

Figure 5.8. The United Nations World Tourism Organization (UNWTO).

Source: Image by Flickr

International tourism was impacted by the global economic and social crisis, but is expected to increase significantly over the following ten years, according to an ILO report1 prepared for the Forum. By 2019, the World Tourism Organization (UNWTO) projects that the sector will support 296 million jobs worldwide.

After several years of growth, the tourism industry experienced a decline that started in the second half of 2008 and worsened in 2009A severe recession in economic activity in the sector occurred globally as a result of a sharp decline in tourist flows, duration of stays, and spending, as well as tighter regulations on business travel expenses.

Foreign visitor arrivals were one of the worst hit by the recession, falling by 4% in 2009, and it was anticipated that international tourism revenues would fall by 6% by the end of 2009. The Middle East (-4.9%), Europe (-5.7%), and the Americas were the geographical areas most severely impacted by the drop in global international travel (–4.6 per cent). Based on a relatively limited volume of travel, only Africa shown consistent growth (+2.9% each year). According to the research, global employment in the tourist sector grew by around 1% between 2008 and 2009 despite the financial crisis. However, there were large regional variations in how the crisis affected employment in the hotel and restaurant industries. Employment in the Americas fell by 1.7%, but increased by 4.6% in the Asia and Pacific region, demonstrating the resilience of the labor market.

5.8.1. A Need for More Social Dialog

It was noted that the demanding work environment in the tourism industry also increases the value of social dialog in the workplace, and that when such procedures are formalized, they actually create opportunities for fruitful collaboration within significant companies in the hotel and tourism sector.

Participants also remarked that it is challenging to implement a universal and formalized social conversation due to the importance of small and medium-sized firms (SMEs) within the sector.

It has become clear that there is a need for innovative and creative social conversation forms that are in line with the reality in a small business in the industry.

According to Wolfgang Weinz, ILO Senior Technical Specialist for the Hotels, Catering, and Tourism sector of the Sectoral Activities Department, this could take the shape of regular consultation between owners/managers and workforce representatives as well as surveys of employee satisfaction and attitudes to highlight worker concerns.

5.8.2. Training and Skills Needs

The Forum gave particular attention to education and vocational training as necessary prerequisites for the operational effectiveness of the industry. It has become clear that its staff members need more training and education, particularly in the areas of health, because they often lack professional expertise. According to Wolfgang Weinz, "the increasing importance of technology, more demanding clients, as well as environmental concerns, require a more effective client/employee interaction. The amount of skill, professionalism, dedication, fervor, devotion, and soft skills of the workforce determine industry competition and productivity."

Participants concurred that talented individuals who are willing to stick with their organization are those who are highly motivated. Some of the soft qualities required, such as civility, discipline, conscientiousness, self-assurance, adaptability, inventiveness, and timeliness, can be improved through training. The management skills of the industry's management were also found to have gaps.

"We are now experiencing a talent gap in our workforce. The staff at hotels needs training. The International Hotel and Restaurant Association's President, Ghassan Aidi, said that it was the only way to raise the standard (IHRA).

5.8.3. The Role of Governments

In order to expand the tourism business, the role of the government has been deemed essential. A site needs a variety of services, such as infrastructure and good destination marketing, to become a desirable travel destination.

Hotels and restaurants also have a huge potential to reduce poverty in many parts of the world. To promote better coordination and advantages for local communities, partnerships between the public and commercial sectors should be strengthened.

Neb Samouth, a panelist from the Cambodian government and participant in the discussion, claimed that "community-based tourism and ecotourism have helped over 30 local communities, offering alternative forms of income and employment. Local initiatives support the preservation of natural resources and serve as an effective forum for social interaction."

The need to improve working conditions and the important role of social dialog were stressed by Ron Oswald, General Secretary of the International Union of Food, Agricultural, Hotel, Restaurant, Catering, Tobacco and Allied Workers' Associations (IUF): "Depending on what kind of tourism and of employment tourism brings to the poorest parts of the world, it can contribute to poverty reduction.

Workers in tourism need to be given the ability to raise themselves out of poverty through representation organizations and ultimately, in ILO terms, collective bargaining."

5.9. CHALLENGES OF EMPLOYMENTS IN TOURISM AND HOSPITALITY

Unequal treatment: All types of employees are not given the same treatment. For instance, gender imbalance is seen in the sectors, according to ILO-UNDP and Thomas. Women produce 50% of the world's food and perform 66% of the labor, but they only make 10% of the money and control 1% of the property.

Such conditions exist because women have less access to land, capital, and education than males do. Women often labor at home or in family businesses without legal protection, and they experience discrimination and overwork in both their professional and personal lives.

According to the UNWTO report on the assessment of the opportunities and challenges that women face with regard to employment in tourism conducted in 2011, unskilled or semi-skilled women tend to work in the most vulnerable jobs, where they are more likely to experience poor working conditions, inequality of opportunity and treatment, violence, exploitation, stress, and sexual harassment.

Additional barriers to women's economic empowerment in developing nations include low levels of education and training, pervasive poverty, poor maternal health, and a lack of sex education, in addition to socio-cultural barriers.

Businesses may be situated some distance from residential regions, especially in less wealthy nations and towns, which places a burden on women's time and travel budgets since they typically have limited access to resources for both time and money.

The other issue is that men predominate in key managerial roles while women are more likely to work at lower levels and in jobs with limited chances for career advancement (Vargas, N.d). In managerial and senior roles, women are underrepresented.

For instance, women make up only 32% of managers in EU-based businesses, 10% of members of the management boards of the largest corporations, and 29% of scientists and engineers in Europe.

Family and caregiving obligations are still not distributed fairly. Women are mostly responsible for taking care of dependent family members. Women opt to take parental leave at a far higher rate than men.

Women are frequently pulled out of the workforce due to this circumstance, as well as the lack of facilities for child care and elder care: the employment rate for women with dependent children is only 62.4%, compared to 91.4% for males with dependent children.

Vergas had also mentioned that, in contrast to other businesses, employment in the hotel, catering, and tourism sectors tends to be more focused on those under the age of 35. In Spain 43.4 per cent of workers in the sector are aged 25–34.

Swarbrooke had noted that employers frequently reject female applicants for employment who are beyond a specific age (generally a 35-year-old restriction) and want images of female applicants. In terms of hiring and promoting persons with disabilities, the tourism sector likewise does poorly.

Poor pay and working condition: According to a report from the European Foundation for the Improvement of Living and Working Conditions (Eurofound), the standard payment rate in hotels and restaurants is low when compared to the standard salary in practically all EU member states. The tourism and hospitality industries are known for their low hourly wages, long workweeks of 50 hours, few to no breaks during peak season, and overtime work without additional pay.

Inappropriate management practices, corporate philosophies, and leadership styles, such as unplanned hiring, paying little attention to staff turnover, hiring foreign workers, viewing employees as costs rather than assets, rigid leadership that runs counter to the idea of democracy, insufficient training, and numerous monotonous tasks, like using computers to make reservations and operating theme park rides, are among the other challenges.

The reason for this is that employment in the tourism industry is frequently threatened by one or more of the following factors: seasonality (during some months of the year, there are relatively few tourists, thus the tourism and hospitality industries reduce staff).

Part-time and/or excessive hours of work; Low-paid (or unpaid) family labor; and informal or occasionally illegal labor, where measurement is noticeably more challenging. A very good example is tourists visiting Ethiopia between September and February, with their number significantly decreasing during the other months.

Furthermore, the industry's various challenging circumstances have an impact on employment chances in tourism and hospitality.

For instance, terrorist attacks are occurring in a variety of tourist hotspots, like ISIS, Bokhara, Alishavave, etc. new infectious illness prevalence, including the so-called Ebola and Zika viruses; the application of e-business, e-marketing, and virtual tourism; and global economic problems in the tourist and hospitality industries.

Many industries in the tourist and hospitality are losing customers as a result. This is the main cause of firms terminating employees, which raises the unemployment rate.

5.10. FUTURE OF EMPLOYMENT IN THE HOSPITALITY INDUSTRY

Since last year, many individuals have questioned whether the hospitality sector has fully recovered. The pursuit of hospitality careers and education by both professionals and students has declined, and some have even switched to the healthcare industry.

On the basis of the gradual rise they observed beginning in the summer of 2020, experts assert that employment in the hospitality industry should resume with full force in the ensuing few years. In fact, they worry that a talent shortage—which was already a problem before the pandemic began—might arise after the business recovers.

Now that 2020 has passed, there's no need to search farther if you're wondering what the employment forecast is for the hotel industry. You can learn important information about the hotel industry's future in this article. Additionally, you will discover the most in-demand positions, trends, and talents in the hospitality industry.

5.10.1. Will the Hospitality Industry Recover?

Despite the fact that limits are continually shifting and currently getting stricter, we have good reason to think the hotel sector will rebound. This theory is based on what we have seen at a worldwide level thus far, namely that once restrictions are eased, people start to travel and venture forth once more.

Travelers now feel more secure and protected because to the higher standards of hygiene, and establishments like hotels, restaurants, and pubs are getting better at putting safety and security precautions into place quickly for both employees and customers.

On the other hand, the introduction of vaccinations also appears encouraging for the revival of the industry. Despite this, the hotel industry will not stay the same (along with most other industries). Events with a hybrid format, Smartphone reservations, wellness declarations, and thermal scanning could all be included.

5.10.2. Is Hospitality Still a Good Career Path?

Even though there are a lot of things that will affect how the hospitality industry develops in the future, it will still be a fulfilling career choice. This is because the essence of and main motivators for this industry will not change, e.g.,

- Meeting people from different cultures
- Working in a diverse and dynamic environment
- Travel opportunities

You can exploit the difficulties posed by the pandemic to develop creative answers and methods if you decide to study hospitality at this time.

The most recent innovations and academic studies on how to handle market crises will also be available to you through your hospitality education in 2021.

If you already work in hospitality, this is a fantastic moment to hone your resilience and leadership abilities. Along with the transferable skills you acquired during your study, your capacity for change adaptation will be another benefit for your profession.

There is no doubt that you will need to put more effort into your own learning and development if you want to stand out as a hospitality student, professional, or jobseeker in this day and age. Since a result, you should be aware of the trends that are influencing the hospitality sector as they have a direct impact on the jobs, education, and skills required in the field.

5.10.3. What Is the Job Outlook For Hospitality?

The industry with the best outlook for the labor market, according to Michael Stull, senior vice president of the international employment business ManpowerGroup North America, is leisure and hospitality, at 26%.

Although his prediction is geared at the American market, it may be quite similar for other places that are well-known in the hospitality and leisure industries.

Our own research indicates the top 10 departments with more open jobs, i.e.:

1. Sales and Marketing
2. F&B Kitchen
3. F&B Service
4. Management
5. Housekeeping
6. Front office
7. Spa Wellness
8. Customer Service
9. Concierge
10. Rooms Division

The top 8 active job categories that were opened on Hosco and published in the past three months, are:

1. Cook
2. Spa Wellness (Beauty advisor)
3. Housekeeping
4. Server
5. Guest Relation
6. Sales
7. Bartender
8. Night Auditors

5.10.4. Trends Influencing Hospitality Jobs

Although many businesses are adapting their business models through innovation, the human touch and personalization in hospitality continue to be crucial elements. In this regard, front-office positions in the hotel industry, like those of receptionists, concierges, and other staff members, are being re-imagined and updated to reflect technological improvements.

Due to the diversification of hospitality occupations, other roles could come with greater responsibility. For instance, self-paced, online learning has become increasingly common in hospitality education, which reflects these trends. Leading hospitality schools in particular point to several common patterns for forthcoming jobs, i.e.:

1. **Technology:**

   Instead of just performing traditional tasks, future hospitality occupations will also require increased expertise in application management and troubleshooting. Before 2020, hotels and resorts began experimenting with artificial intelligence and virtual reality to enhance their operations and guest experiences. Marriott Hotels, for instance, had put in place "the Teleporter."

   The first robot concierge was tried by Hilton, in contrast. Contactless check-in and QR code menus helped to speed this trend during the pandemic, and it is expected to continue.

2. **Unique Experiences:**

   Individualization and the creation of enduring experiences will rule. Due to the economic crisis, travelers are more cost-conscious than ever and favor customized experiences over pre-packaged getaways.

3. **Sustainability:**

   Jobs like "Sustainability Analysts" and "Asset Managers" in hotels and resorts will see tremendous growth as a result of the increased attention being paid to environmental health. Additionally, as people look to reconnect with nature, more people will travel to neighboring tourist attractions, cultural landmarks, and national treasures.

5.11. CONCLUSION

This chapter summarizes all about employability and skills in tourism and hospitality. This chapter also explains vocational employability skills, graduate employment, and Types of Employment: SME, MNC, Non-Profit, Public, and Self, etc. There are various roles available in the tourism and hospitality sector so this chapter also explains the employment opportunities with all the essential skills needed for a career in tourism and hospitality.

There are a lot of challenges and issues of employment in tourism and hospitality; it is the major issue in tourism and hospitality. So, in this chapter, Challenges of employment in tourism and hospitality are discussed.

Towards the end of the chapter, it also provides highlights on the future of employment in the hospitality industry by answering some questions like will the hospitality industry recover? is hospitality still a good career path? What is the job outlook for hospitality? After answering these questions, trends influencing hospitality jobs are discussed.

REFERENCES

1. 10 Essential Skills for a Career in Tourism and Hospitality, (2022). CareerAddict. Available at: https://www.careeraddict.com/top-10-skills-needed-for-a-job-in-hospitality (accessed on 14 June 2023).
2. 5 Essential Soft Skills for Careers in Hospitality and Tourism, (n.d). Topuniversities. Available at: https://www.topuniversities.com/student-info/careers-advice/5-essential-soft-skills-careers-hospitality-tourism (accessed on 14 June 2023).
3. Aynalem, S., Birhanu, K., & Tesefay, S., (-1). Journal of Tourism & Hospitality. Longdom Publishing S. L. Available at: https://www.longdom.org/open-access/employment-opportunities-and-challenges-in-tourism-and-hospitality-sectors-15201.html (accessed on 14 June 2023).
4. Employability Skills – university of Exeter, (n.d). Exeter. www.exeter.ac.uk. Available at: https://www.exeter.ac.uk/ambassadors/HESTEM/resources/General/STEMNET%20Employability%20skills%20guide.pdf (accessed on 14 June 2023).
5. Employability Skills: The Skills you Need to Get a Job, (n.d). SkillsYouNeed. Available at: https://www.skillsyouneed.com/general/employability-skills.html (accessed on 14 June 2023).
6. Employment in the Tourism Industry to Grow Significantly, (2011). www.ilo.org. Available at: https://www.ilo.org/global/publications/world-of-work-magazine/articles/WCMS_157893/lang--en/index.htm (accessed on 14 June 2023).
7. Firth, M., (2019). The Employability and Skills Handbook for Tourism, Hospitality and Events Management Students. Routledge. academia. Available at: https://www.academia.edu/40082114/The_Employability_and_Skills_Handbook_for_Tourism_Hospitality_and_Events_Management_Students (accessed on 14 June 2023).
8. What is the Future of Employment in the Hospitality Industry? (2021). Hosco. advice.hosco.com. Available at: https://advice.hosco.com/en/future-of-employment-in-the-hospitality-industry/ (accessed on 14 June 2023).

Chapter

6 MOBILITIES, TOURISM, AND TRAVEL BEHAVIOR

Contents

6.1. Introduction: Muslim Travel Behavior in Halal Tourism 140

6.2. Cultural Tourism and Event Tourism: Characteristics and Specificities 144

6.3. Theoretical Background and Hypothesis Development 146

6.4. What Is Long-Haul Travel? 152

6.5. Long-Haul Tourists 153

6.6. Travel Motivations 153

6.7. Analysis of the Spatial Range and Influencing Factors of Travel Behaviors 155

6.8. Cruise Traffic in the Mediterranean Area 157

6.9. Approaches and Procedures for Data Achievement 160

6.10. Dynamics of the Colombian Medical Tourism 162

6.11. Systemic Competitiveness of Associative Character: A Theoretical Approach 164

6.12. Conclusion 166

References 168

In the chapter of mobilities, tourism and travel behavior addresses about the cultural tourism and event tourism: characteristics and specificities. This chapter also mentions the theoretical background and hypothesis development. The chapter also highlights the long haul travel, long haul tourists.

The chapter also address the analysis of the spatial range and influencing factors of travel behavior. At the end of the chapter, this chapter provides the highlights on the cruise traffic in the Mediterranean area, approaches and procedures for data achievement and dynamics of the Colombian medical tourism.

6.1. INTRODUCTION: MUSLIM TRAVEL BEHAVIOR IN HALAL TOURISM

In order to establish Halal tourism infrastructure and facilities, Halal travel packages, and Halal travel activities, industry participants, marketers and policy makers must have a thorough understanding of Muslim tourist behavior.

In order to draw in Muslim tourists, it will also assist them in developing targeted marketing messaging. Muslim customers are one of the fastest-growing market segments, and their needs cannot be ignored by destination marketers and tourism operators, despite the fact that the past or future size of the Muslim population in the world does not indicate how much of this population actively practices its religions or how religious its members are or will be.

According to the State of the Global Islamic Economy study, which was created by Thomson Reuters in partnership with DinarStandard, the global Muslim travel industry was worth $140 billion in 2013, or 11.5% of all global spending.

The same analysis projects that this market would generate $238 billion in revenue in 2019 and account for 13% of all spending worldwide. UAE and Singapore are at the top of the Halal Travel Indicator, which measures the health of a nation's family-friendly/Halal travel ecosystem in relation to its size.

From the standpoints of both business and academic research, there is currently a growing interest in Halal tourism. Many emphasize the necessity of using Islamic teachings and principles as the foundation for any plan to produce or sell Halal tourism goods and services.

The concept of Halal tourism is defined by a number of phrases, each of which offers a recommended policy to help elucidate the situation. Building on this foundation, this chapter makes the claim that the terms "Halal," "Islamic," and "Muslim-friendly" have different meanings when used in relation to tourism.

It is believed that a better grasp of the aforementioned concepts may allay the worries of numerous researchers and marketers regarding Muslim tourists' Halal tourism behavior.

6.1.1. Is it Halal or Islamic?

Finding the appropriate terminologies and ways to explain topics is a problem that both academics and practitioners must deal with. Currently, the terms "Halal tourism" and "Islamic tourism" are most frequently used.

There are now misunderstandings about the two terminologies due to the multidisciplinary nature of the issue. The phrases are therefore frequently used interchangeably by academics in both conceptual and empirical works as if the two notions are comparable.

However, as the Quran distinguishes between "Mu'minoon" (an "Islamic person") and "Muslim" (those who claim to be Muslims), it is important to understand the distinction between the terms "Islamic," "Halal," and "Islam related."

It is important to remember that Islam is the name of the religion, whose first prophet was Adam, and whose last prophet was Muhammad. "A condition of peace gained through devotion to God" is one of the definitions of Islam. Islam also means "submission" in other languages. As a result, the term "Muslim" is used to refer to someone who practices Islam and submits to God. According to Sheikh Yusuf al-Qaradawi, a renowned Islamic scholar and the head of the International Union of Muslim Scholars, the definition of Halal is "That which is permitted, with respect to which there are no restrictions, and the doing of which the law-Giver, Allah, is allowed" in one very authoritative book titled The Lawful and the Prohibited in Islam.

As a result, the term "halal" refers to something that is permitted under Islamic law. Halal is one of the five deeds (al-ahkam al-khamsah) that Muslims use to classify the morality of human conduct, along with Fard (obligatory), Mustahabb (preferred), Makruh (undesirable), and Haram.

From an Islamic viewpoint, Halal, as described above, refers to any tourism practice or activity that is "permissible" in accordance with Islamic teaching. Because of this, referring to anything as "Halal" or "Sharia compliant" has the same connotation.

The term "Islamic" is strictly used to describe only that which is directly related to the faith and its doctrines (such as Islamic law/Sharia, Islamic values, principles, and beliefs, and Islamic worship). As a result, it is more similar to the Arabic word "Mu'minoon."

This is due to the fact that Islam presents the faith as an ideal based on its two main sources, the Quran and the Prophet's Sunnah. The terms "Islamic males" and "Muslim men" have different connotations, and it follows that Muslim men, Muslim women, and Muslim countries may not all be Islamic.

In addition, the Niyyah, or intention, of an activity must be present for it to be considered "Islamic." When someone does an action or activity with the aim of seeking God's pleasure, it is considered to be Islamic and accepted by God. This is based on a well-known hadith attributed to the Prophet Muhammad (Figure 6.1):

'Verily actions are by intentions, and for every person is what he intended. So the one who's "hijrah" (migration) was to Allah and His Messenger, then his "hijrah" was to Allah and His Messenger. And the one whose "hijrah" was for the world to gain from it, or a woman to marry her, then his "hijrah" was for what he made "hijrah" for' (Al-Bukhārī and Muslim).

Figure 6.1. Halal friendly destination.

Source: Image by Flickr

6.1.2. Halal Tourism and Islamic Tourism Terminology

The UNWTO defines tourism as "the activities of people visiting and residing in places outside their regular environment for not more than one consecutive year for leisure, business, and other objectives."

Tourism was defined by Cook et al. as "the temporary migration of individuals to locations other than their usual places of employment and abode, the activities engaged in during their stay in such locations, and the facilities established to accommodate to their needs."

From the above description, it is clear that tourism encompasses both supply and demand. The demand side of tourism involves people traveling, whether they are Muslim or not, and the supply side comprises activities and amenities at tourist destinations that cater to their demands.

The key takeaway is that in order to understand the behavior of Muslim visitors, marketers must first comprehend the demand side, or the demands of Muslim tourists in the context of Halal tourism.

In the literature on travel and destination marketing, some scholars have attempted to define Islamic and Halal tourism. Most people, however, have not considered the Sharia, the target audience (i.e., Muslims or non-Muslims), the location of the activity (i.e., Muslim vs. non-Muslim countries), the product and service given (i.e., food, facilities), and the purpose of travel (Figure 6.2).

Figure 6.2. Islamic Centre.

Source: Image by Wikimedia Commons

Most of these definitions are ambiguous and applied to several contexts. In other words, Islamic tourism and halal travel are viewed as equivalent ideas. Islamic tourism, for instance, is described as "the encouraging of tourists likely to fit the standards of Sharia law" by Jafari and Scott.

The definition ignores the tourists' religion (Muslims) and other factors in favor of focusing on Islamic law and its requirements to satisfy tourist needs. We believe that "Halal tourism" rather than "Islamic tourism" is better described by Jafari and Scott.

6.2. CULTURAL TOURISM AND EVENT TOURISM: CHARACTERISTICS AND SPECIFICITIES

The idea and concept of cultural tourism and event tourism are described in the analysis that follows. Music-related events are presented as examples.

6.2.1. Cultural Tourism as a Platform for Event Tourism

Tourism that is inspired by culture is referred to as cultural tourism. The term "culture" refers to the entirety of humankind's moral and spiritual values, as well as its material and intangible possessions. A cultural attraction within a particular destination is necessary to promote and develop cultural tourism. The World Tourism Organization (WTO) describes cultural tourism as a type of travel that "includes all the elements of the journey through which travelers learn about the mindset and lives of those they meet during their stay.

"Cultural tourism is defined by the Association for Tourism and Leisure Education (ATLAS) as "the movement of people to cultural sites that are remote from their place of residence, with the goal to gather information and new experiences in order to meet their cultural requirements."

According to World Tourism Organization specialists, cultural tourism includes:

Travel for young people that emphasizes cultural and educational elements, such as study abroad programs, excursions to learn a second language, and international summer camps;

- exchanges of artists, writers, and the events celebrating this exchanges;
- rural tourism, holidays to the families of the visited region/ country;

- holidays in the villages integrated in the cultural local environment;
- attendance at religious festivals, national, and international festivals;
- cultural circuits, which includes visits to museums, historical sites, and attending various cultural events;
- Pilgrimages to historical and religious places and monuments.

Cultural tourism differs from other forms of travel in a variety of ways due to the nature of its offerings, the way it manifests itself and its changes over time. Moreover, cultural tourism is positioned at the intersection between tourism and culture, borrowing some of the characteristics from both of the two phenomena.

According to this, the particularities of cultural tourism are:

- Due to the particular services that are included, such as those provided by cultural institutions (museums, cathedrals and monasteries, festivals, performances, numerous cultural events, etc.), cultural tourism has a multifaceted content.
- cultural tourism targets a clientele with a certain level of culture and education or which is being educated;
- cultural tourism is generally a more expensive product, but not a luxury one;
- compared with other forms of tourism, cultural tourism seasonality is less accentuated (Figure 6.3).

Figure 6.3. Pilgrimages and Santiago Museum.

Source: Image by Flickr

6.2.2. Event as a Tourist Attraction and the Specificities of Event Tourism

The word "attraction" refers to the force that draws something to it, the force of attraction, or something that piques people's attention. Tourist attractions, in Kuen's words, "are the fundamental resources of any tourist destination that shape and condition the tourist offer and the development of tourism in general."

Their concept is determined by the requirements, motives, and actions of tourists. Regardless of whether they are a part of a particular location or have a strictly geographically defined occurrence, all tourist attractions are significantly spatially marked (as the authors underline).

Frequently, the word "event" connotes an attraction. The words "happening, phenomenon, consequence, and competition" are all part of the English noun event. The Croatian language has adopted this word and uses it in that way.

The word "event" does not appear in the Croatian foreign word dictionary, indicating that it was only used after the 1990s. An event can be a ceremony, a sporting event, a congress, a conference, an exhibition, a festival, or a party.

It can also be something that occurs or is perceived to occur, or an occurrence, especially one of significant importance. It is a phenomenon that can occur in several contexts. The concepts of event management and event organization are frequently associated with the word event.

The literature particular to the four main categories of events and associated venues (i.e., business, sport, festivals, and entertainment) that rule practice and have drawn the most attention from scholars is then examined while arguing the categorization of events using a thematic approach.

6.3. THEORETICAL BACKGROUND AND HYPOTHESIS DEVELOPMENT

6.3.1. Theoretical Background

6.3.1.1. Tourist Satisfaction

According to specialized literature, a visitor's estimate of satisfaction following a variety of experiences in a location is based on their subjective viewpoint. In this sense, visitor pleasure with a destination essentially reflects what the visitor anticipates.

In other words, it is the accomplishment of an expectation that arises from lived experiences in the location and might be articulated as an emotive state, a cognitive state, or a combination of both.

According to this viewpoint, it is a favorable evaluation of the benefits attained in the location by the visitor, or a compilation of judgments about the quality the visitor perceives.

6.3.1.2. *Functional Benefit Perceived by the Tourist*

It has been asserted that a person who is solution-focused tends to select the option that best meets their needs. When choosing a destination in this context and when presented with various options, the functional benefit that the traveler perceives is crucial.

This advantage, which may be characterized as the utility attained through functional value, is produced by meeting an individual's cognitive requirement as a result of their adaptation to the environment. The tourist accomplishes this through enjoying the location's attributes.

The perceived functional advantage will be higher the more a place meets the practical needs of visitors. The stronger the impact on the visitor, the more effectively you can meet their various needs. As a result, the tourist destination's useful attributes will increase the likelihood that people will choose to visit it.

6.3.1.3. *Hedonic Benefit Perceived by the Tourist*

The satisfaction of the experience in terms of service, linked to their fantasies and sentiments, can be described as the tourist's perception of the hedonic benefit. From this perspective, the tourist's perceived hedonic benefit is important while choosing a destination. It will be used to gauge how much pleasure you got from a certain location. Psycho-sensory experiences, which emerge from the desire for stimulation and stimulation, are what cause this. This is a way to depict multimodal experiences that are triggered by sensory stimuli. In this sense, tourist destinations are made up of a variety of goods and services that provide stimulating sensory experiences for visitors. The visitor will appreciate the trip more if they have better multisensory experiences. Visual interaction with the terrain will actually be a crucial component because tourism destinations are primarily esthetic.

6.3.1.4. *Symbolic Benefit Perceived by the Tourist*

The symbolic benefit experienced by the tourist is defined as the advantage gained through the various components of self-concept, which, in large part, results from the evaluation of others, whether real or imagined, and consists of different representations by each person, linked to a specific set of social circumstances.

According to this viewpoint, self-concept is divided into four categories: actual, ideal, social, and ideal social. An individual's self-image is referred to as their "actual self-image," their "ideal self-image," their "social self," and their "ideal social self," which is how they would like to be regarded by other significant individuals.

In order to strengthen or improve their self-concept, travelers could change how they behave while visiting a place. However, they could also go to locations that have special significance to them or others. The tourist will thus be able to appreciate the sense of belonging that the destination offers through the symbolic benefit they perceive.

6.3.1.5. *Cognitive Perception of the Tourist Destination*

It has been argued in specialized tourism literature that cognition can be defined as both the collection of people's beliefs, ideas, and impressions about a specific place, on the one hand, and as objective knowledge of a place or the quality of the physical characteristics of the environment, on the other.

In a similar line, a study of the literature on tourism indicates the diversity of studies connected to tourists' cognition in terms of defining a location and an image.

As a discretionary activity with more options and less restrictions than the majority of other human undertakings, tourism is viewed from this angle as having relevance for tourist behavior.

More specifically, cognitive perception refers to the processes involved in paying attention to stimuli and events, comprehending them, recalling past experiences, making judgments, and making purchasing decisions.

6.3.1.6. *Affective Evaluation of the Tourist Destination*

In terms of tourism, a tourist's affective evaluation of a location can be described as their perception of the location's affective qualities or how they feel about it.

As a result, tourists will be inspired by the esthetics of travel places and their attractions as well as by their emotional characteristics, which will help them meet their psychological demands. A destination's relationship with its guests creates affective associations that may be favorable, neutral, or negative.

According to this viewpoint, such a connection between products and feelings causes the buyer to have an image of the product in their head. Therefore, an affective evaluation can come about in one of two ways: either through intrinsic sensory/motor processes that produce primitive or imperfect affective assessments, or through a conceptual outline and transformation.

6.3.2. Hypothesis Development

6.3.2.1. Functional Benefit Perceived—Tourist Satisfaction

In terms of tourism, the perceived functional advantage has a favorable impact on visitor satisfaction. Similar to this, consumer behavior literature points out that features that serve as functional advantages raise consumer happiness and can be incorporated into a product's design and esthetic.

The practical advantage a visitor receives from a location's features may also increase visitor satisfaction. From this perspective, the following theory is put up in relation to tourism destinations:

H1. Tourist pleasure is directly impacted favorably by the functional benefit experienced through the tourist attraction.

6.3.2.2. Hedonic Benefit Perceived: Tourist Satisfaction

The hedonic advantage experienced in terms of tourism has a favorable impact on visitor satisfaction. Hedonic events may boost a customer's sense of satisfaction, according to that consumer. Customers with hedonic tendencies seek out enjoyable activities that boost enjoyment. The enjoyment of leisure experiences influences tourist satisfaction.

In a similar manner, the contentment of visitors may be influenced by how much they appreciate a location's attractions. In light of this, the following theory is put up in relation to travel destinations:

H2. Satisfaction among visitors is directly influenced by the hedonic benefit received from the tourist destination.

6.3.2.3. Symbolic Benefit Perceived: Tourist Satisfaction

When it comes to tourism, the symbolic benefit felt has a favorable impact on visitor satisfaction. The amount of happiness is directly and favorably related to the symbolic value the tourist receives from their visit to a particular location.

Positive behavior in relation to a place is consequently impacted by this positive effect. According to tourism literature, the symbolic benefit for the visitor may greatly raise their degree of happiness with the site.

Consequently, the connection between the symbolic advantage and satisfaction can serve to improve how people view the provision of hospitality services. The following theory is proposed in the context of tourism sites along the same lines:

H3. Tourist satisfaction is directly influenced favorably by the symbolic benefit experienced through the tourist destination.

6.3.2.4. Cognitive Perception of the Tourist Destination: Functional Benefit Perceived

To get the desired practical benefit, the visitor needs to have a good cognitive perception of the location. From a marketing perspective, it has been noted that the functional benefit is primarily instrumental and cognitive in nature for the customer.

Similarly, research on consumer behavior has shown that the functional gain can be attained as a reaction to a cognitive demand. In this regard, the tourist's ability to get practical benefits could be enhanced by having a positive cognitive assessment of the location's attributes and features. In the context of tourism sites, the following theory is proposed from this angle:

H4. The cognitive perspective of a tourist destination directly influences the functional benefit that the visitor perceives.

6.3.2.5. Cognitive Perception of the Tourist Destination: Hedonic Benefit Perceived

It has been suggested that there is a direct connection between cognitive perception and hedonic benefit when it comes to consumers who seek experiences (high degrees of stimulation).

As a matter of fact, although while the hedonic benefit is obtained through the consumer's experience of the product's and/or service's most ethereal qualities, it is a result of the cognitive perception of the most tangible aspects of the experience.

According to this theory, good cognitive perception will favorably increase the consumer's hedonic benefit. In this regard, the following theory is put out in relation to tourism destinations:

H5. The cognitive perception of a tourist destination directly influences the hedonic benefit that the visitor experiences.

6.3.2.6. Cognitive Perception of the Tourist Destination: Symbolic Benefit Perceived

Based on the evaluation of its attributes, a favorable cognitive perception of the location is crucial to having a positive impact on the tourist's symbolic advantage. Literature demonstrates that the symbolic benefit that the buyer perceives may be cognitive in character.

To be able to reproduce each self-concept using what is already known as well as what is unknown, the tourist's cognitive perception of the location will be important in this regard. A tourist's positive cognitive evaluation of the location will therefore encourage the anticipated symbolic benefit. Within the context of tourism attractions, the following theory is put out from this angle:

H6. The symbolic benefit that a traveler perceives is directly influenced by their cognitive perception of a particular destination.

6.3.2.7. Affective Evaluation of the Tourist Destination: Functional Benefit Perceived

The consumer's perception of the functional advantage might serve as a tangible expression of the affective judgment. A product's emotional attributes have a favorable impact on the functional value that the consumer perceives.

The customer's perception of the functional benefit of a product is increased when it is emotionally appealing. The following theory is put out in the context of tourist locations in a similar manner:

H7. The practical value that a tourist perceives is directly influenced by how they feel about a particular tourist site.

6.3.2.8. Affective Evaluation of the Tourist Destination: Hedonic Benefit Perceived

The tourist will be able to experience the desired hedonic advantage if they have a positive affective rating of the location. It has been stated that the affective assessment directly affects any potential hedonic gain that the consumer may experience when making a purchasing decision.

In this sense, feelings are the primary mechanisms by which hedonic value for the consumer is produced. To achieve the hedonic benefit that travelers anticipate when visiting a destination, it is essential to understand the affective basis of consumer experiences.

Therefore, in order to achieve the hedonic advantage that the tourist seeks, a favorable affective rating of a location will be beneficial. In the context of tourism attractions, it is suggested that the following hypothesis be true:

H8. The tourist's perception of the hedonic advantage is directly influenced by their emotive assessment of a particular travel place.

6.3.2.9. Affective Evaluation of the Tourist Destination: Symbolic Benefit Perceived

The emotive judgment is highly important, according to the research, because it precedes the symbolic benefit. It has been stated that there is a strong correlation between the individual's subjective affective assessment and the symbolic gain they perceive. In order for the consumer to receive the apparent symbolic gain, the emotive appraisal is essential. In light of this, the following theory is put up in relation to tourism destinations:

H9. The symbolic benefit that a traveler perceives is directly influenced by how they feel about a particular destination.

6.4. WHAT IS LONG-HAUL TRAVEL?

Traveling across a vast distance or over a long period of time is referred to as long-haul or long-distance. According to the Dateline project, long-distance travel entails a minimum 100 kilometer excursion.

According to Boerjan, long-distance air travel takes more than five hours or a journey of more than three thousand miles. Long-haul travel, according to Letho et al., is defined as "travel by plane for more than four nights outside of the international region."

Long-haul, multi-destination excursions, and the usage of airplanes were uncommon in the past. Travel is now regarded as a habit, and for some people, it is an integral part of their existence. Traveling to a place is today more convenient than ever because to the introduction of new technology (such as new airports, direct flights, and others), which have made it possible for people to travel farther and more comfortably.

6.5. LONG-HAUL TOURISTS

There are differences between long- and short-distance travelers in terms of vacations, excursions, finances, lodging, and other factors. According to Tidelwall, Faulkner, Hwang, and Fasenmaier, the majority of long-distance travelers are independent, take multi-destination trips, and want to fulfill a variety of travel objectives.

For long-distance travelers, who typically take more time gathering information before their trips, the pre-holiday phase is crucial. Long-haul travelers, according to McKercher and du Cros, are more likely to be open to experiencing a different culture and have a propensity for longer journeys.

In addition to viewing their excursions as uncommon, frequently once-in-a-lifetime occurrences, long-haul travelers regard them as opportunities for learning and personal growth. According to Bao and McKercher, there are certain parallels between long-distance travelers and backpackers, including their independence, tight budgets, preference for basic lodging, and mode of travel.

However, according to Crouch, travelers who travel great distances are less price sensitive and more likely to spend more since they value product quality and features.

6.6. TRAVEL MOTIVATIONS

According to Myers, motivation is the need or desire that spurs behavior toward a goal. According to Hsu et al., tourists act in certain ways for specific reasons. Tourists typically travel for a variety of reasons, including relaxation, prestige, nostalgia, bettering a relationship, social contact, adventure, health, personal growth, novelty, education, isolation, and romance.

According to the fundamental idea of motivation, a dynamic process of internal psychological components (needs, desires, and goals) will result in a degree of discomfort and tension, and this will satisfy the criteria.

According to several academics, a destination's push and pull elements influence tourists' decisions to vacation there. Individuals are motivated to travel by internal causes known as "push factors," whereas external influences known as "pull factors" affect where, when, and how they travel.

According to Uysal and Hagan, push and pull factors are functioning independently, and tourists travel because their decisions to do so are driven by a variety of interests. According to Bashar, the majority of push factors that are origin-related are intangible or internal to the individual tourist.

However, motivation studies vary depending on the context (i.e., place and time). In addition, every tourist is unique based on their age, motivation, and needs. According to studies, the "knowledge" and "culture" of Hong Kong was the driving force behind long-haul tourists.

According to a research by Correia et al. on Portuguese tourists visiting Latin America and Africa, long-distance tourists primary push reason was to "relax," while their primary pull motivation was "landscapes."

According to a research by Assiouras et al. in Greece, long-distance tourists from China, Korea, and Japan had "culture and history" as their primary pull factors, while "knowledge" was their primary push element (Figure 6.4).

Figure 6.4. Parachute Adventure.

Source: Image by Max Pixel

6.7. ANALYSIS OF THE SPATIAL RANGE AND INFLUENCING FACTORS OF TRAVEL BEHAVIORS

To examine the influence of internal subjective elements and external objective stimuli, the spatial scale of WHMs' travel behavior can be investigated.

6.7.1. Analysis of Spatial Range of Travel Behavior

6.7.1.1. Short-Distance Travel Behavior

Tourist activities typically allude to long-distance travel habits. But "recreational activities" are nevertheless generally necessary in daily living. Some WHMs noted that they would spend their free time participating in a variety of common outdoor recreational pursuits, like going for walks, taking pictures, cycling, going on outings, dining out, and so forth.

Their goal was to relax. According to one participant, "In the afternoon, everyone might meet up for nearby sightseeing, cycling somewhere, or taking photography nearby, or go out for a walk at night to enjoy a sky full of stars.

To me, this also represented enjoyment (J3, 2014). Going shopping was also a way for some WHMs to appreciate the novelty and exoticness of the shopping experience. For instance, one participant said, "At first, even going to the grocery store was exciting.

I found going to the grocery store to be quite interesting (J2, 2014). Another person added, "I really liked going to the local market to buy strange stuff and to see new people, activities, and objects. You'll feel differently if you fit in with their lives (C2, 2014).

In essence, WHMs were more inclined to participate in regular recreational activities and broaden their personal horizons when they were exposed to a fresh environment. Additionally, cultural experience events allowed WHMs to connect with locals and successfully integrate into the community, allowing them to learn more about the local way of life and culture.

For instance, with reference to their personal encounters with regional norms, one participant shared, "Once, I was a visitor at a Japanese household. We performed New Year's Day rituals at a shrine, including currency offerings, lotteries, and bell-roping for happiness (J3, 2014).

Another WHM who visited Canada said, "I was glad to spend Thanksgiving at my friend's place, and we had a turkey dinner." I was finally able to experience the foreigner feeling" (C4, 2014). Thus, it seems clear that engaging in regional traditions and events gave WHMs a richer travel experience.

Furthermore, social events were the most popular leisure activities among WHMs. WHMs were frequently drawn to these social meetings because there was no requirement for a certain time, place, or format, making it simple to meet travelers from all over the world.

As an illustration, one participant said, "The farm owner enjoyed planning gatherings. There was one almost every week, offering food, entertainment, and socializing opportunities (J3, 2014). Consequently, social events were among the finest ways for WHMs to make friends abroad, increase their contacts, and strengthen their bonds with one another.

Another member added, "Everyone explained how to prepare and consume specific cuisines. The foreigners particularly enjoyed my specialty dish of salt and pepper chicken (A1, 2014).

6.7.1.2. Long-Distance Travel Behaviors

The tourist resources in each country vary in attractiveness due to geographical and historical variances. This gave WHMs a variety of tourist-related options, including the chance to visit unusual natural wonders and take part in festivals or themed activities.

Terrain, climate, ecology, and cultural landscapes are all special attractions, such as unique fauna, unusual surface kinds and structures, and cultural and historical landscapes.

As stated by one of the participants, "We went to a spot called Mount Rainier. It was stunning to see the glaciers. As far as the eye could see, there was glacial terrain (C4, 2014). We ascended Uluru, another participant added. It's not unexpected that the locals refer to it as a sacred mountain because it changes color depending on the time of day when it receives sunlight (A1, 2014).

In addition to seeing these natural sceneries up close that are not present in Taiwan, WHMs emphasized that seeing animals up close was a unique tourist experience, such as being able to view polar bears, elks, and brown bears in person. According to one participant, "I spotted a brown bear and it was really enormous near the entrance of the tourist center.

When you approached, it was standing directly next to you. The anticipation was high (C2, 2014). Another participant said, "I even held a koala." This participant had traveled to Australia where she had the opportunity to touch a koala and take pictures with kangaroos. I believe Taiwan won't allow us to accomplish this. Kangaroos were also used in several of the photos. They were very tall (A1, 2014).

Other tourist destinations that drew WHMs were residences, shrines, ancient cities, canals, cathedrals, and other special cultural and historical settings. One of the participant remarked, "Do you know a place named Gassho Village? There are numerous little thatched and wooden cottages there.

It like a fairy-tale. It was a recognized World Heritage Site. National landmarks like Matsumoto Castle and Torii were also present. I went to several intriguing shrines (J2, 2014). WHMs had the chance to engage in cultural experiences and a variety of entertainment thanks to special festivals and specialized events.

They encourage WHMs to learn more about the history and culture of their community. Consequently, they are quite alluring, and as a result, many travelers make the special trip to visit them, as is the case with St. Patrick's Day in Canada and the Obon Festival in Japan.

A participant said, "Every year around August 15, the Japanese celebrate the Obon Festival, which features the Bon dance. I had a girl's costume on. I'll never forget it; it was a very interesting experience (J3, 2014).

Moreover, if WHMs are sports enthusiasts, they will make special arrangements to attend significant sporting events, such as the Australian Open tennis tournament, Canadian ice hockey, American professional basketball games, and so forth.

One participant remarked, "My dream was to see the Australian Open and I truly went! Tennis is a sport I enjoy watching, but watching it live is different from watching it on television (A1, 2014). Another participant added, "Those who play basketball when they come here (North America), they must witness the NBA. You feel high playing the games (C2, 2014).

6.8. CRUISE TRAFFIC IN THE MEDITERRANEAN AREA

When viewed from a worldwide perspective, the cruise market is constantly expanding, fiercely competitive, and investing in new technologies every year. According to CLIA data, 24.2 million cruise passengers traveled the world in 2016, and 25.3 million are anticipated worldwide in 2017.

This global phenomena resembles the market patterns in the Mediterranean Sea, where the number of passengers climbed from 24.2 to 34.3 million between 2007 and 2016 to 10.1 million in less than 10 years. At the start of 2017, there were 26.3 million lower berths, an increase of approximately 80% since 2007.

Even while the near future does not appear to continue the same growing trend, the Mediterranean is currently the second-most popular cruise destination, after the Caribbean, as indicated in the introduction. Actually, a more fair assessment needs to account for the strong growth of other locations, particularly those in Asia.

More than 100 ports welcome cruise ships throughout the Mediterranean Sea. Of course, there are significant differences in accessibility, dimensions, facilities provided, and closeness to tourist attractions, infrastructures, investments, and local laws.

It is important to note in this analysis how cruise trips are impacted by the particular geopolitical environment in several Mediterranean regions. This is one of the key arguments for why it would be preferable to divide the Mediterranean region into macro regions and/or distinct seas.

It is clear from both tables that the biggest and most favorable data variation between 2015 and 2016 is for the Adriatic region. When compared to the rest of the Mediterranean and the Black Sea, this is very obvious. This last and the East Mediterranean region have a relatively significant variety, but in negative terms.

As can be seen from the data presented above and when taking the last five years into account, the number of passengers in the entire Mediterranean region has climbed to 2.4%, but the overall number of calls has significantly fallen to more than 8%.

The increasing size of ships provides a simple explanation for this statistics; in fact, over the same period of time, passenger movements per call have increased by 13%.

With more than 5 million passenger movements, the Adriatic Sea has more than 30 cruise ports that are mostly known for transit traffic. In this case, Venice serves as the home port and is the top travel destination for the entire region, accounting for 31.7% of passengers in 2016 among the 20 most significant local ports.

Dubrovnik comes in second with 833,588 passenger movements (16.5% of the total) followed by Corfu with 748,916 passenger movements (14.8% of the total) in order of popularity. Igoumenitsa, with 7623 passenger trips (or 0.2% of all passenger movements), is the final destination on the list.

It is noteworthy that five Italian ports are listed among the top 20 Adriatic Sea ports. Venice, Bari, Trieste, Ancona, and Ravenna are in order of significance. With 2.3 million cruise passengers, Italy has the most passenger movements overall.

This statistic equates to exactly 44.6% of passenger movements in the Adriatic, or close to half of all passenger movements in the region. It must be noted that numerous ports have opened or been expanded in Italy during the past few years, and more are scheduled to become operational soon.

With 1200 additional berths between those that have just opened and those that are scheduled to open in the next year or two, the Veneto region was the one most interested in this procedure.

On the one hand, the aforementioned information demonstrates how this region may be essential to the growth of Italy's tourism industry as a whole, as well as that of the maritime tourism industry specifically.

On the other hand, the region's uniqueness and the data from the passengers demand the implementation of sustainable development strategies. Venice is a clear representation of a conundrum. Traffic limitations are currently in place and a progressive control is necessary for the most popular site, which also has a highly delicate ecological.

The need for improved collective actions has significantly increased over the past few years, and it is not hard to locate research studies and programs that stress the necessity of intergovernmental cooperation at all levels linked to local business associations and stakeholders.

The Adriatic marine tourism destinations might be coordinated as part of a comprehensive policy that includes works to protect the water and its ports of call as a solution to this issue. An option for resolving this could be to apply the most recent and unique aggregation model from Italy across the entire region, as it is one of the primary participants in this scenario.

The model is a type of business network contract, also known as a Contratto di Rete d'Impresa. In Italy, where the tourism industry is particularly successful, this concept is used extensively. This type of activity would unify the region as a single travel destination and increase coordination across various nations, deals, brands, and marketing activities.

Additionally, the corporate network agreement might serve as the direct counterpart of the European Union bodies, advocating for more creative growth plans and environmental protection laws.

The network's territorial reach, the number of its affiliates, the variety of legal entities it includes—such as NGO, Ltd., foundations, and local administrative bodies—as well as its coordination system as embodied by the management body make it easy to see the network as a valuable and influential partner. The members of a network contract, known as a Retisti, currently operate as a limited business under a common name, governed by an external management body, and sharing a bank fund while maintaining their individual identities (Figure 6.5).

Figure 6.5. West Mediterranean sea areas.

Source: Image by Wikimedia Commons

6.9. APPROACHES AND PROCEDURES FOR DATA ACHIEVEMENT

The research's exploratory methodology tested the viability of using online conversation analysis to define the meaning of sustainable tourism as it emerged from web users, including policymakers, in the Adriatic-Ionian region.

For this purpose, the definition of the area has been taken from official European publications. Four popular tourist locations in the area that share similar visitor appeal and size have been selected.

The selection of specific locations was based on factors like past performance, prior notoriety, and the intensity of the tourism industry relative to the local population and geographic area. As a result, despite the fact that each of these locations has received a Blue Flag, Cesenatico, San Benedetto del Tronto, and Makarska and Split in Croatia have been recognized as the most representative locations.

The volume of themed messages that could be retrieved from Twitter for each of them was gathered during the previous seasonal peak period, which ran from July 15 to August 31, 2016. The meaning of sustainable tourism as it emerges from the texting of actual actors can be determined by text analysis of tweets.

It was decided that a well-known eco-label tool's definition of sustainable tourism would serve as a good substitute. A few sketches are provided below and the Foundation for Environmental Education (FEEBlue)'s Flag program for marinas and beaches have been selected for this.

The Blue Flag program was established in France in 1985, and South Africa joined in 2001. It has been in operation in Europe since 1987 and outside of it since then. Today, Blue Flag is a truly international program, with participation from an ever-growing number of nations:

"The Blue Flag program promotes sustainable development in freshwater and marine areas. It challenges local authorities and beach operators to achieve high standards in the four categories of: water quality, environmental management, environmental education, and safety."

As stated in the preceding paragraph, the decision is appropriately in line with the goal of incorporating sustainable tourism into the marketing discourse. It seems reasonable to view the Blue Flag program as a reliable sign of communicated sustainable tourism and the aforementioned four categories as criteria for choosing keywords to be detected.

On the side of social media, the rationale for choosing Twitter is the following: This online platform has 313 million users, 1 billion monthly accesses to websites that embed tweets, 82% of mobile users, 3860 employees, and 35 offices worldwide, as well as 79% of accounts registered outside of the US, 40 languages spoken, and 40% of technicians on the staff as a whole.

Even though Twitter is thought to be in a falling phase right now, it still has the potential to aggregate content created on other platforms, such as Instagram, Facebook, and Tumblr, and it still permits, albeit with certain restrictions, the survey of user-generated data.

6.10. DYNAMICS OF THE COLOMBIAN MEDICAL TOURISM

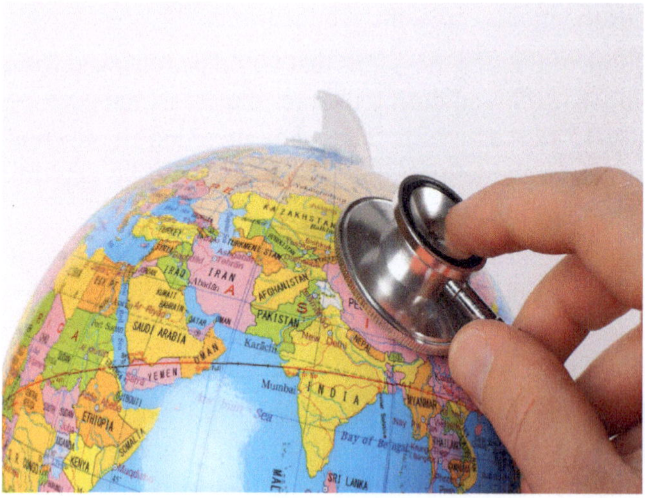

Figure 6.6. Doctor with stethoscope and globe in his hand. Medical network and health care in Europe and Africa, concept of world medicine, epidemic control, medical tourism.

Source: Image by Flickr

The medical tourism industry, which is still in its infancy, has about 100,833 patients in the beginning of 2016. According to the Ministry of Industry and Commerce, approximately 10,117 patients sought medical treatment in Colombia, while 2351 went there for cosmetic operations and preventative medication. However, the wellness procedures are the ones that attracted most patients with a total of 88,405 in 2016 (Figure 6.6).

The need for medical services is driven by a desire for treatments that are less expensive than in industrialized nations like the United States, which is Colombia's top importer of patients.

According to the Productive Transformation Program, between 2013 and 2016, there was a decrease in demand for medical services by 42.3% as a result of a variety of factors, including exchange rate volatility and clinics' insufficient readiness to receive patients in accordance with internationally recognized protocols.

The most popular medical operations are braces, pelvic floor reconstruction, heart surgery, breast augmentation, liposuction, and non-invasive weight loss regimens.

The demand for noninvasive medical procedures is anticipated to rise abroad for a number of reasons, according to McKinsey & Company. The first is that more tourists are seeking for unique opportunities to learn more about the places they visit.

The second is that direct investments are less in this area compared to those in curative and preventive medicine, where infrastructure is required for extremely sophisticated procedures with medical and technology equipment.

This category has considerably better potential for growth because it doesn't require as much investment (although more publicity and marketing are needed to promote this category).

Compared to 2013, investments in clinical, hospital, and esthetic infrastructure rose by 120% in 2015. The investments have been distinguished by offering a high-quality service, mostly to international patients looking for treatment and interventions (invasive or noninvasive), whether for curative, preventative, esthetic, or well-being goals.

The procurement of specialist medical equipment and the training of human resources received the majority of the infrastructure investment in private clinics and highly complicated health centers. The tax structure of free economic zones, which lowers fixed costs, was used to fund a significant portion of this investment in health infrastructure.

The World Development Indicators report that the country's coverage of healthcare services has significantly increased thanks to incentives for investment in the sector.

The total construction area of health centers (hospitals of high complexity) in Colombia's seven largest cities, including Bogota, Medellin, Cali, Barranquilla, Bucaramanga, Pereira, and Armenia, has increased steadily since 2005, according to the National Administrative Department of Statistics of Colombia.

By 2020, investments in health infrastructure are anticipated to grow moderately but more slowly. This is because health services, which are mostly geared toward patients from the United States and Panama, have stabilized and consolidated.

These people are increasingly interested in wellness and esthetic medicine services rather than highly complex curative and preventative medicine.

The industry's international positioning strategy includes patient loyalty (mainly from foreigners and Colombians abroad, though national patients with high purchasing power are not disregarded), and preference for national medical procedures that combine interventions, esthetic, and wellness treatments with the opportunity to learn about Colombia's tourist attractions.

The development of the sector goes hand in hand with the accreditation of top-tier clinics internationally as a means of luring foreign patients and increasing strategic alliances with foreign providers of medical services, such as insurance firms, guilds of the international healthcare sector, and research and development that might boost the reputation of Colombian doctors abroad.

In this regard, four clinics with accreditation from the Joint Commission International and the Global Healthcare Accreditation Program in medical procedures of high complexity serve as a resource for international patients and insurance companies to send their customers to receive medical operations.

According to De la Puente, Colombia's market share in the globe will rise from 1.4 to 5% in 2020 under the present medical tourism business model, with patients seeking esthetic procedures and naturally occurring weight loss being the most in-demand services. It is anticipated that the top five patient-producing nations will remain the United States, Canada, Aruba, the Antilles, and Panama. The majority of them are Colombians who choose to use medical services in their home country because of the high costs of treatments in the country where they reside.

6.11. SYSTEMIC COMPETITIVENESS OF ASSOCIATIVE CHARACTER: A THEORETICAL APPROACH

When there is still misunderstanding and ambiguity surrounding the concept of competitiveness, it is challenging to research it as a paradigm, leaving a gap in both theoretical and applied clarity.

The Competitive Advantage of Nations by Porter is a first conceptual approach to the idea of competitiveness as a goal in and of itself based on national enterprises' ability to maintain and increase their participation in global markets while simultaneously raising the population's standard of living.

The diamond model, which is characterized by the allocation of four factors that dramatically increase an industry's participation in global markets, is used to further clarify the description:

(1) the conditions of factors,

(2) the context for the strategy and business rivalry,

(3) the conditions of demand, and

(4) The related industries and support.

The government's role as an exogenous agency, which should have a limited impact on local businesses' competition for foreign market share, is the main criticism of Porter's model. Although the inclusion of the government component suggests a better exogenous condition for businesses to grow their market share, there is no research of how the four elements affect the government in its capacity as a regulator and supporter of the businesses' competitiveness.

On the other hand, Rugman and Cruz's Diamond Model, adds the government component to Porter's original diamond as a result of the increasingly competitive environments in which technological differentiation of both processes and the use of digital media strengthen the output products.

It manifests a more participatory approach in the creation of spaces that increases market share for national companies based on complementary elements.

Jones relates the idea of competitiveness to the commercial performance of businesses with various legal forms, with the growth of their assets and enhancements to the welfare of the population (quality of life, employment, and impact on the environment), vertical integration, innovation, efficient resource management, incorporation of new technology, decentralization of production, etc., relating the three levels of competitiveness: business competitiveness, national competitiveness, and industrial competitiveness.

According to De la Puente, the distribution of goods and services determines corporate competitiveness, but the creation of an environment that promotes industrial competitiveness determines national competitiveness.

In other words, the government is no longer seen by Porter and Rugman as an indirect supporter of global trade, but rather as a regulating agent and a facilitator of economies of scale.

Contrary to the conventional viewpoint on the general concept of competitiveness mentioned above, wherein the government only acts as a regulator, facilitator, and promoter of the expansion of market niches for domestic companies focused on favorable macroeconomic conditions (which visualizes competitiveness as a result of the conditions of "macro" variables and their influence on the costs of production and factor prices), The systemic competitiveness approach takes into account both institutionalism and neolaborism, which emphasize the relevance of institutional factors in management, innovation, and learning as well as the importance of labor for a competitive advantage.

This paradigm is built on four levels: micro, meso, macro, and meta. Systemic competitiveness is the outcome of complex and dynamic interaction patterns between a nation's government, businesses, intermediary institutions, and the organizational ability of a society.

Systemic competitiveness is defined as the understanding that the success of an industrial development is not only due to a production function at the micro level or to stable macroeconomic conditions, but also to government initiatives and policies that promote competitiveness in businesses or organizations from the confluence of an environment that encourages social, judicial, political, and macroeconomic stability, incorporating sociocultural adaptations that promote social competition (target level).

The Organization for Economic Co-operation and Development (OECD) report developing the OECD Competitiveness Indicators Platform states that these strategies were compiled and systematized into a holistic concept known as "structural competitiveness."

Using its tactics for brand positioning, economies of scale, and customer loyalty, the government has increased its involvement in the manufacturing and selling of goods and services.

6.12. CONCLUSION

In this chapter, the knowledge of Muslim tourist behavior in the context of Halal tourism is pertinent to industry players, marketers, and policy makers alike in developing Halal tourism infrastructure and facilities, Halal travel packages and Halal travel activities.

Towards the end, this chapter concludes theoretical background and hypothesis development with the Long-haul and short-haul tourists are different based on holidays, trips, budgets, accommodation and others.

REFERENCES

1. Asero, V., & Skonieczny, S., (2018). Cruise tourism and sustainability in the Mediterranean. Destination Venice. *Mobilities, Tourism and Travel Behavior – Contexts and Boundaries* [Preprint]. Available at: https://doi.org/10.5772/intechopen.71459 (accessed on 14 June 2023).
2. Battour, M., (2018). Muslim travel behavior in halal tourism. *Mobilities, Tourism and Travel Behavior – Contexts and Boundaries* [Preprint]. Available at: https://doi.org/10.5772/intechopen.70370 (accessed on 14 June 2023).
3. Corinto, G. L., & Curzi, F., (2018). Analysis of online conversations for giving sense to sustainable tourism in the Adriatic-Ionian region. *Mobilities, Tourism and Travel Behavior – Contexts and Boundaries* [Preprint]. Available at: https://doi.org/10.5772/intechopen.7037.
4. Foris, D., Popescu, M., & Foris, T., (2018). A comprehensive review of the quality approach in tourism. *Mobilities, Tourism and Travel Behavior – Contexts and Boundaries* [Preprint]. Available at: https://doi.org/10.5772/intechopen.70494.
5. Krajnović, A., & Gortan-Carlin, I. P., (2018). Music event as a tourist product: Specifics, issues, challenges. *Mobilities, Tourism and Travel Behavior – Contexts and Boundaries* [Preprint]. Available at: https://doi.org/10.5772/intechopen.71338.
6. Marinao, E., (2018). Determinants of satisfaction with the tourist destination. *Mobilities, Tourism and Travel Behavior – Contexts and Boundaries* [Preprint]. Available at: https://doi.org/10.5772/intechopen.70343.
7. Ni, C.-Cheng., Tsao, C.-Yu., & Wang, Y.-hsiang., (2018). The international decision-making and travel behavior of graduates participating in working holiday. *Mobilities, Tourism and Travel Behavior – Contexts and Boundaries* [Preprint]. Available at: https://doi.org/10.5772/intechopen.70331.
8. Pacheco, M. A., (2018). Systematic competitiveness in Colombian medical tourism: An Examination. *Mobilities, Tourism and Travel Behavior – Contexts and Boundaries* [Preprint]. Available at: https://doi.org/10.5772/intechopen.70135.

9. Shahrin, N., & Marzuki, A., (2018). Long-haul travel motivation by international tourist to Penang. *Mobilities, Tourism and Travel Behavior – Contexts and Boundaries* [Preprint]. Available at: https://doi.org/10.5772/intechopen.72411.

Chapter

7

CONTEMPORARY DEVELOPMENTS AND PERSPECTIVES IN INTERNATIONAL HEALTH SECURITY

Contents

7.1. Introduction 172

7.2. Expanding the Definition and the Scope of Health Security 173

7.3. Emerging Infectious Diseases 174

7.4. Chronic Health Conditions and Access to Care 175

7.5. Trauma Systems 177

7.6. Emergency Medical Services 179

7.7. What Is Ransomware? 182

7.8. International Health Security: Potentially Harmful Aspects of Social Media Use 185

7.9. Human Trafficking 188

7.10. Theories of Morality 189

7.11. Health and Safety at Work 194

7.12. Advantages of Microfluidic Lab-on-a-Chip Technology 196

7.13. Conclusion 198

References 199

In the chapter of contemporary developments and perspectives in international health security it talks about the scope of health security. This chapter also explain the emerging infectious disease, chronic health conditions and access to care. The chapter provide highlights on the trauma systems and ransomware.

This chapter also helps to understand emergency medical services, international health security: potentially harmful aspects of social media use. The chapter also addresses about the human trafficking, theories of mortality, health and safety at work and advantages of microfluidic lab on a chip technology.

7.1. INTRODUCTION

The field of international health security (IHS), which falls within the broader category of health sciences, is complicated and extremely heterogeneous. Within that broad context, IHS includes all areas that could potentially affect (or, more specifically, potentially endanger) the state of human health and wellness, from natural disasters to outbreaks and pandemics, including rapid urbanization, population growth, various environmental issues, crucial supply chain vulnerabilities, misuse of antibiotics leading to unpredictable microbial resistance patterns, and even false information on social media (SM).

The realization that although clearly dominant factors within the overall context of health security, subjects like emerging infectious diseases (EID), food, water, and pharmaceutical supply chain safety, medical and health information cybersecurity, and bioterrorism constitute only a small portion of forces able to actively modulate the wellbeing and health of human populations led to the need for this expanded definition of health security.

7.1.1. Why Did We Embark on This Project?

IHS experts have access to a wide range of materials, and many of the literature sources are distinguished by very high content quality. Why then, one might ask, embark on another book project on the topic? There are several significant justifications.

The first and most important factor is the limited scope of the scientific literature that has already been published on this subject, which is getting more and more complicated. IHS is now a far broader and complex field, with new topics and subdomains that are fast developing and are quite detailed. According to several experts on the subject, Planetary Health is the upcoming conceptual advancement.

The authors and editors of this collection of chapters believe that IHS, when used broadly, inherently encompasses a variety of concepts, such as "public health," "global health security," and "planetary health," while also ensuring that the overall context of its implementation maintains compatibility with the modern understanding of the world as a collection of independent countries and states that operate within a well-defined set of international conventions.

The authors hope to inform and energize readers to become more aware and active in issues related to health security at local, regional, national, international, as well as global scales by covering new topics and offering a variety of solutions and approaches on rapidly emerging infectious, social, environmental, and bio political dilemmas, etc.

7.2. EXPANDING THE DEFINITION AND THE SCOPE OF HEALTH SECURITY

The standard definition of "global health security," or GHS, which largely focuses on bioterrorism and EIDs, is rather constrained. We are essentially talking about human security, though in a far broader sense. Since everyone and everything in the world is interconnected, we must coordinate our efforts, identify common ground, and work as a team to combat the problems we face.

As the Ebola virus, the Zika virus, and the Coronavirus Disease 2019 (COVID-19) have shown, a hazard to one (person, region, country, or continent) can swiftly develop into a threat to all. Here, we discuss relevant currents that have an impact on all cultures from various views, experiences, and facts.

Fundamentally, the Editors hope that the readers will identify health security gaps that they may not have previously seen as well as opportunities that they may take on in their own communities, nations, and continents.

The current book broadens the more conventional and constrained understanding of IHS to include fresh and significant elements of climate change, global health, cyber security, social determinants of health, technology/information management, and several other closely connected fields.

7.3. EMERGING INFECTIOUS DISEASES

Pathogen emergence (and reemergence) poses a serious hazard to public health, affecting both high-income regions (HIRs) and low- and middle-income countries (LMIRs). It is difficult to identify new IHS risks in this field, mostly because new infections and pathogens tend to originate in a non-specific manner and frequently without warning.

Potential EIDs can result from diseases that are zoonotic, vector-borne, food-borne, antibiotic-resistant, or prone to outbreaks, as well as from a lack of access to clean water and sanitary facilities. These illnesses, in addition to causing fatalities, can have a considerable negative impact on the local economy and strain its health care infrastructure. Future responses to new infectious illnesses can be made better by using the knowledge gained from previous outbreaks.

For instance, the well-documented response to the Ebola outbreak in 2014 demonstrated both a vulnerability to this crucial IHS danger and severe inefficiencies in the existing global public health infrastructure.

The initial reaction lacked organization and coordination. The tracking of cases was also insufficient because the case trajectory ostensibly declined before the actual global transmission took place. Furthermore, transmission was able to spread at an accelerating rate because of the overstretched nature of local health care systems, the tardiness of communication within the surveillance system in place, the rapid urbanization of the region, and the prevalence of poverty where people lacked access to adequate water and waste management infrastructure.

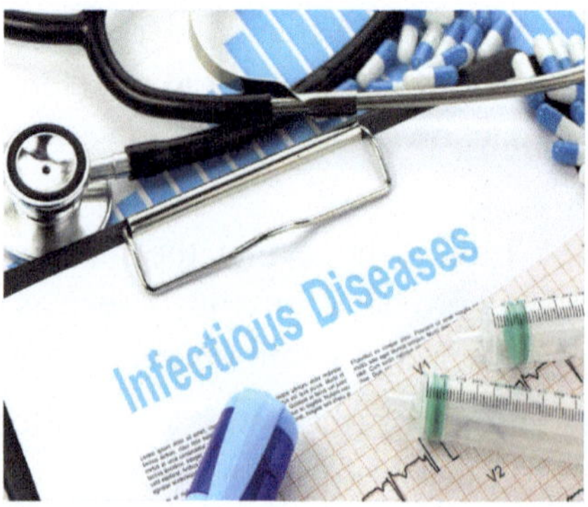

Figure 7.1. Infectious Disease.

Source: Image by Picpedia

Finally, local disputes may make any public health and medical activities either extremely risky or maybe impossible, even when there are sufficient resources available to address EID issues (Figure 7.1).

The appropriate steps must be taken in order to stop the emergence or reemergence of potentially fatal diseases. Among these strategies is active surveillance for (timely response to) epidemics. Prioritization must be given to issues like antibiotic stewardship, vector control, increased efforts to fight poverty, improved water and waste management, and the necessity of education and the transmission of critical information to all stakeholders.

In order to effectively combat EIDs, it is essential to coordinate all activities as a single, global front. IHS places a great deal of weight on issues like mass migration and advancing travel convenience in a world that is characterized by constant globalization.

Collectively, the aforementioned characteristics may speed up the spread of some diseases, and when combined with insufficient reaction, lack of detection, and a lack of awareness of the many risks involved, local outbreaks can quickly turn into pandemics.

Finally, it will be necessary to provide assurances that any corresponding public health initiatives will receive sufficient funding across all domains and services as the international community starts to broaden its collective focus to include some of the more prevalent chronic, communicable, and non-communicable diseases. Additionally, institutional processes will need to be built to guarantee ongoing evaluation, training, and readiness to stop the rise of global complacency.

7.4. CHRONIC HEALTH CONDITIONS AND ACCESS TO CARE

Chronic health disorders (CHCs), such as diabetes and depression, are getting more and more prevalent as the world's population ages, which is posing a serious threat to public health. This is especially true when it comes to healthcare access across LMIR. The number of people over 60 in the world is rising, and by 2100, that population is projected to have tripled, from 962 million in 2017.

A significant proportion of this expanding population group will come from the LMIR of today, and most of this expansion has been credited to better medical care, which has led to longer life expectancies. In addition, as a result of lower birth rates, the world population is aging. The combination of an aging population and CHCs must be seen as a threat to global security, especially as the resource-to-patient ratios start to fall.

The most prevalent CHCs affecting this growing elderly population are dementia, heart disease, diabetes, obesity, mental health disorders, stroke, HIV, sexually transmitted diseases (STDs), malnutrition, sensory impairments, substance abuse, polypharmacy, bladder irregularities, and mobility issues, with LMIRs facing the greatest difficulties (Figure 7.2).

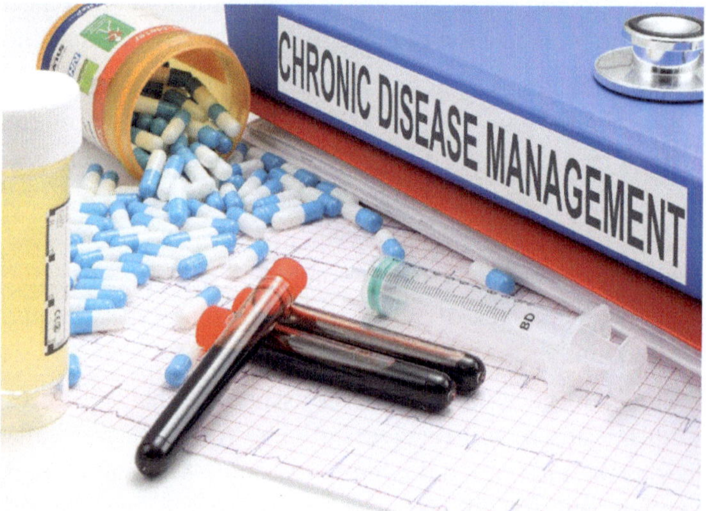

Figure 7.2. Chronic health and access to care.

Source: Image by Blue diamond gallery

External circumstances could make these hurdles even more difficult to overcome. It will be necessary to find innovative solutions to competing forces like rising healthcare expenditures and a physician/provider shortage.

Technology, an innovation-driven emphasis, and the development of global economies of scale can all be used in long-term sustainability strategies. Examples in detail include telemedicine, community/family participation, preventative measures, and at-home primary care.

Home health and telemedicine initiatives will need to replace less effective models where transportation to and from medical institutions was the norm due to the difficulty of accessing care in many LMIRs. Community education initiatives can assist in educating patients and family members about aging in place care.

Most of the globe now has access to "smart phone" technology because of quick advancements in telecommunication, and alternatives for telemedicine based on this technology can offer portable care in locations where traditional health-care information infrastructure is insufficient or out-of-date.

Point-of-care (POC) technology enables the initiation of improved preventative interventions, such as diabetes and cardiovascular screening, as well as behavioral changes that enhance wellbeing and continuity of care.

Community involvement and committed non-health care degree employees educated in particular interventions will be beneficial for increasing the current health care workforce. Similar to this, offering senior citizens work or volunteer options helps keep them involved, active, and invested in their local communities.

The elderly LMIR populations served by CHCs can gain from patient-centered treatment in addition to the operational solutions mentioned above. In order to assure sustainability, it will be crucial to reduce the total level of disability among the elderly. Promoting personal independence, socializing, and mobility can be achieved by providing appropriate dental care, hearing aids, glasses, mobility aids, and activities for groups that are appropriate.

Patients can have a proper diet while still being involved in the community by receiving nutrition support from nearby farmland linked to food banks, receiving meals from food programs, or participating in food sharing programs that involve group meals.

Early discussions about end-of-life (EOL) care are crucial from a variety of angles and perspectives because they can help balance patient and community goals while reducing pointless, expensive EOL treatments.

Patient experiences can be enhanced by using home hospice, moral EOL policy initiatives, and increased community education. CHCs in the elderly LMIR population must be handled pro-actively and systematically.

To effectively treat the growing aging population in LMIRs, health systems must emphasize goals of care that improve quality of life, increase independence while reducing handicap, and create safe, secure access to health-care in an innovative, technology manner.

7.5. TRAUMA SYSTEMS

Traumatic injury is a disease that knows no borders; it is one of the main causes of morbidity and mortality worldwide and has a disproportionately large negative impact on resource-poor nations. In 2016, 1.4 million individuals were killed by road accidents alone, with men and boys making up the majority (74%) of the victims.

Morbidity and mortality linked to trauma continue to be a major global health concern despite increased knowledge of injury causes and prevention due to the expanding global population, traffic, and urbanization.

An organized network of prehospital emergency care and a standardized system of trauma care that can be replicated and provided to patients in rural community hospitals and significant academic tertiary care facilities alike are necessary for ensuring timely access to advanced trauma care as an international health security measure.

The Advanced Trauma Life Support (ATLS) course has been one of the biggest efforts to date to improve trauma care. The American College of Surgeons created this training program in 1978 in response to the terrible incident of an orthopedic physician driving his jet who crashed into a Nebraska cornfield with his family, killing his wife and seriously injuring his three children.

This surgeon described the deficiencies in the emergency medical system and demanded systemic improvement to enhance the care of trauma victims worldwide. Given that 30% of all trauma deaths happen within 60 minutes of damage, ATLS places special emphasis on the first stabilization and resuscitation of the trauma patient, dubbed the "Golden Hour."

Although there is little information on the impact of ATLS on trauma mortality, the research that is available supports its use as a strategy of reducing mortality and enhancing global healthcare systems.

More than 90% of injury-related fatalities take place in low-income nations, where there are few preventative initiatives, urgent care facilities, and resuscitation centers equipped to provide quick, sophisticated resuscitation.

The majority of these deaths are caused by road traffic injuries. In order to reduce global morbidity and mortality from trauma, early advanced trauma treatment must be established. This is done in part by disseminating trauma education programs. According to studies, as the number of professionals with ATSL training rises, the rates of deaths that may have been prevented or at least theoretically prevented decline. One study comparing trauma-related fatalities before and after the introduction of ATLS and other specialized trauma education courses in the Rwandan capital discovered a significant decline in the mortality of badly injured patients in the six months after their start. After the implementation of a primary trauma care course, practitioners from nations throughout East, Central, and Southern Africa managed trauma patients more effectively, according to another study. The implementation of ATLS in the Netherlands led to a noticeably better trauma outcome within the first hour of admission.

The classification of trauma centers led to improvements, according to a study on the effect of required ATLS training on care processes in rural America. After receiving ATLS training in Trinidad and Tobago, better trauma patient outcomes have also been noted.

ATLS has undergone numerous iterations since it was first developed. Since the middle of the 1990s, over 1 million physicians have been trained in its concepts, which are now considered standard of care in more than 50 nations throughout the world. With proven outcomes in terms of better patient outcomes, care practices, and training, ATLS has grown into a global resuscitation program.

7.6. EMERGENCY MEDICAL SERVICES

An emergency medical service can be defined as a comprehensive system that arranges for staff, facilities, and equipment to give health and safety services to victims of unexpected illness or injury in an efficient, coordinated, and timely manner. In order to avoid needless fatality or long-term morbidity, EMS's mission is to promptly deliver medical care for out-of-hospital medical situations.

Health security is fundamentally based on timely access to healthcare in an emergency. Prompt Prehospital emergency care that is given right away and swift transport of the sick and injured to a medical institution can save lives, lower the likelihood of temporary disability, and significantly improve long-term outcomes.

An essential component of any efficient and effective health care system, according to the World Health Organization, is the emergency medical services (EMS) system. Emergency medical service (EMS) has developed in industrialized countries into a vital link in the chain of survival for those experiencing illness or injury outside of a hospital.

It also makes a substantial contribution to the general efficiency of a healthcare system and the wellbeing of a society. However, the emergence and development of emergency medical services have progressed at a relatively gradual rate and on various timescales all across the world.

Although modern emergency medical services (EMS) were first created by Napoleon to help wounded soldiers, not many significant advancements to EMS occurred until the 1960s. After the Vietnam War, prehospital care and transport concepts were more readily applied from the military to the civilian sector as a result of the increased awareness of care inequities among returning veterans.

By the end of the war, victims of gunshot wounds had a higher chance of surviving in the jungles of Vietnam than in the streets of large American towns. A more structured EMS system was developed in the United States and overseas between 1960 and 1970 as a result of several medical, historical, and social elements coming together.

The Anglo-American and Franco-German models of care, which are still in use today, were two separate forms of emergency healthcare delivery that had diverse ideologues and distinctive characteristics. Although the categorical differences between these two models were more apparent before the 20th century, modern EMS systems all over the world have developed along one of these two frameworks while incorporating different elements of the other.

The Franco-German approach is founded on the "stay and stabilize" attitude, in which the patient is given cutting-edge medical treatment. Physicians in charge of emergency services respond to patients' homes or the scene of an accident to administer advanced medical care. Either the patient is stabilized and given further instructions, or they are taken to the hospital for admittance.

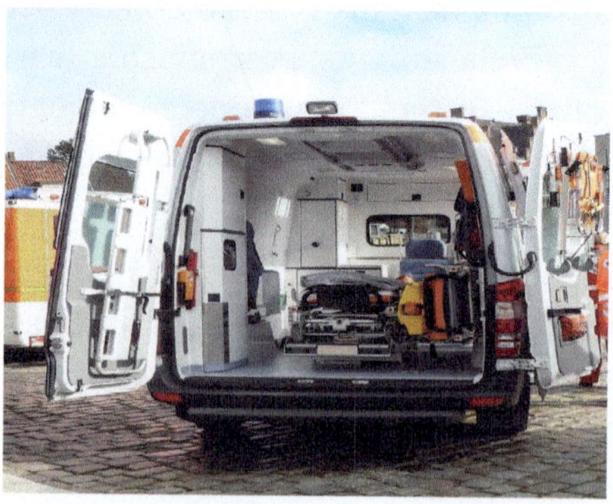

Figure 7.3. Emergency Medical Services.

Source: Image by Needpix.com

As a result, there are fewer EMS transports and less need for emergency room visits and treatment. In continental Europe (e.g., France, Germany, Greece, Malta, and Austria), where emergency medicine is a new field of study, this method of prehospital care—in which EMS functions as an outpatient department of the hospital—is frequently used (Figure 7.3).

The "scoop and run" mentality is the foundation of the Anglo-American paradigm of EMS service. This model's objective is to quickly transport patients to the hospital while minimizing the amount of time spent on prehospital care and interventions.

Instead of public health services or hospitals, EMS services are associated with public safety services in this case, such as fire or police departments. With remote physician supervision, specialized emergency medical technicians (EMTs) and paramedics provide immediate patient care.

In nations where this approach is prevalent, emergency medicine has developed into a well-recognized medical specialty. The majority of the evaluation and treatment for patients begins in emergency departments (EDs). Following that, patients are either released or confined to the wards for more medical care. Examples of countries utilizing this model include the United States, Canada, New Zealand, and Australia.

While all EMS models have a comparable emphasis on stabilizing and transporting seriously ill and injured patients, it is the provision of non-life-threatening treatment that the biggest differences can be discovered.

The Anglo-American model transports the vast majority of patients for examination in an emergency room, in contrast to the Franco-German model, which places more emphasis on treating patients in their homes and avoiding transport if possible. Many studies have compared the patient outcomes and cost-effectiveness of the two approaches because to the stark contrasts between these two systems.

However, because each model is different, the results are challenging to understand. Each has diverse objectives and functions in a unique environment. There is no proof that one model is superior to the other, and the lack of clear standards makes direct comparison difficult.

Unfortunately, many nations around the world still lack the ability to offer the most fundamental level of prehospital emergency care. While EMS practitioners in some nations have expanded their role in responding to medical emergencies and have access to cutting-edge clinical technologies, in others their education and training are significantly more constrained due to a lack of funds, resources, and organizational supervision. No matter how basic or complex a country's prehospital care system is, it is still within their reach to implement the necessary components of a successful model. These components, which have been found to reduce morbidity and death, consist of fast communication and activation of the prehospital emergency response system, instantaneous system response, and concurrent treatment and transport to formal medical care.

It is significant to emphasize that there is minimal proof that the most seriously injured or ill patients benefit from enhanced prehospital interventions. According to studies, the majority of instances may be handled by responders who have simply received training in basic life support.

This has significant ramifications for resource-constrained countries, where cutting-edge life support services and materials could undermine a system by siphoning funds away from less flashy but more practical initiatives that help more people.

The model for developing and managing an EMS system that is best suited to a specific demographic will ultimately be determined by local resources, culture, and values, but it should always be the result of collaboration between local and federal governments with the aim of enhancing health security.

7.7. WHAT IS RANSOMWARE?

In order to conduct an extortion attack, malicious software known as ransomware infiltrates computer systems or other linked devices and encrypts user files. Most frequently, a machine becomes infected with ransomware when a user opens a hacked email or goes to a corrupted website (i.e., drive-by downloads).

Once downloaded, servers (including web and email servers), databases, end-user PCs, and portable media, including private cloud storage services, may also be involved. When someone has access to the encrypted data ("ciphertext") but is unable to decode it to reveal its contents ("plaintext"), that person has violated the privacy intention of encryption.

Asymmetric key and public key are the two varieties of encryption, or cryptography. When using symmetric key cryptography, both the sender and the receiver encrypt and decrypt the data using the same secret key. A pair of keys are used in public key cryptography: a private key and a public key that are shared by both parties (sender and receiver have their own unique private key). Asymmetrical cryptosystem created by combining the two cryptographies is used by ransomware to encrypt data using a randomly generated symmetric key, which is then followed by an encryption using a public key where one party owns the corresponding private key.

In order to decode the data back into "plaintext," the cybercriminal uses the victim's private key to decrypt the symmetric key. The victim then receives the decryption key back, which they can use to restore control of their machine.

Data is rendered inaccessible and unintelligible once it has been encrypted. A pop-up message notifies the user that they must pay a ransom in exchange for the decryption key. Typically, this ransom is paid in an anonymous digital currency like bitcoin.

Until a ransom is paid, ransomware frequently locks up data instead of erasing it. The data may still be encrypted even after the ransomware infection has been eliminated. However, it is crucial to realize that merely infecting a computer with ransomware is insufficient. To obtain an encryption key and report its findings, the ransomware has to contact with a server.

This calls for the use of a server maintained by a business that will overlook any illegal conduct and ensure the attackers' privacy (called Bulletproof Hosting). These businesses frequently have headquarters in China or Russia. Attackers can further mask their own internet protocol (IP) addresses by using a proxy server or virtual private network (VPN) services.

Malware developers' adoption of an intuitive modular design for ransomware dissemination has contributed to an increase in attack counts. This strategy, known as ransomware-as-a-service (RaaS), has grown in popularity. It helps technically unsavvy attackers by disseminating the malware simply through phishing and exploit kits and offers a reliable revenue model.

RaaS is most easily accessed on the Dark Web, where aspiring cybercriminals are given access to an affiliate interface that walks them through the process of getting their ransomware exploit kit and lets them modify settings, choose targets, and choose ransom amounts. Metrics on malware instillations and success rates are also available.

7.7.1. Ransomware Types

Ransomware can be broken down into three categories: crypto, locker, and wipe. Although the two primary categories of ransomware are crypto and locker, newer varieties sometimes include elements from both. The most prevalent type of ransomware encrypts data as well as files. Therefore, if moved to another device, infected files remain unavailable.

Critical system files are often preserved, allowing the device to keep running until the ransom is paid. The greater secrecy of cryptocurrencies is another reason why crypto-ransomware favors bitcoin. However, because of concerns about law enforcement, bitcoin anonymizers and laundering businesses have appeared.

On the other hand, locker-ransomware (a less successful extortion weapon) locks the device by building a digital "locker" around the computer system to prevent access. Contrary to crypto-ransomware, the data stored on the device are usually undamaged and can frequently be recovered by shifting them to another functional computer for access.

Additionally, customers may be able to remotely uninstall the locker-ransomware and forego paying the ransom. However, if remote virus eradication is unsuccessful, ransom payments are often performed using crypto currency or payment voucher systems.

For instance, online betting firms may accept the voucher codes as payment and then transfer the funds to prepaid debit cards. The money is then taken out through money mules.

With the PetrWrap attack, which encrypted the target's master file table (MFT) and forced the operating system (OS) to restart, wipe-ransomware first made an appearance in 2017. Wipe-ransomware, in contrast to crypto- and locker-ransomware, encrypts files and does not allow them to be decrypted even after payment, essentially causing data loss.

7.7.2. Ransom Payment

Online payment options were less accessible before 2005. In order to make money for the attacker, victims were told to transmit ransom payments via checks to offshore accounts, SMS text messages, prepaid cards, or even premium rate telephone numbers.

Due to the fact that they might be tracked, these techniques were dangerous. With the introduction of the relatively anonymous cryptocurrency bitcoin in 2008, ransomware assaults increased. Additional payment options have been made available through the usage of third-party holding corporations like PayPal. Ransomware utilizes dynamic geographical pricing because a person's capacity to pay might differ significantly by location and the local economy. Once a computer or system has been compromised, the ransomware connects to its command-and-control (C&C) server, reports the IP address of the compromised device, and the C&C server responds with a price for the nation connected with that IP address based on a pre-populated database.

Additionally, because there is a higher chance that a ransom will be demanded, fraudsters target corporations more often than individual users. According to reports, a company ransom of roughly $10,000 USD may be ideal because it is both affordable and causes law enforcement to hesitate before opening an investigation.

It's crucial to decide whether to pay the ransom. The FBI of the United States advises against paying ransoms because, in the end, only 50% of victims are able to acquire uncorrupted useable data. The payment of a ransom encourages terrorists to keep targeting healthcare facilities for their attacks.

However, in order to regain data accessibility and prevent additional losses, an estimated 40% of firms decide to pay the ransom. If the hospital lacks business continuity and has a suspect backup, this can happen more frequently.

However, refusing to pay the ransom results in additional expenses like as prolonged outage and recovery, which might cost up to 23 times as much. Smaller organizations have been forced to close after not paying the ransom.

According to the FBI, ransomware-related financial losses exceeded $1 billion USD in 2016 alone, with an average downtime cost per assault of $141,000. The choice of whether to pay the ransom is ultimately one that is made by the individual and is based on the particular circumstances and risks of each instance.

7.8. INTERNATIONAL HEALTH SECURITY: POTENTIALLY HARMFUL ASPECTS OF SOCIAL MEDIA USE

Sharing unreviewed information via social media carries the risk of spreading false information or misinterpretation of such unfiltered content, especially if it is taken out of context. Fact-checking capabilities are outnumbered by "inaccurate or fake news" in the hourly volume of social media messages by a factor of up to 10.

Additionally, the average delay between the dissemination of "fake news" and any "fact-checking" response may be longer than 12 hours, resulting in serious harm before the inaccurate information can be corrected. Therefore, if disinformation is spread by a person who is highly motivated and equipped with the necessary tools, it can have a significant impact on public dialog.

When "false news" is deliberately created and disseminated in the context of IHS, the repercussions may be severe. "Fake news" can be harmful in a number of ways, including by disseminating erroneous assertions (such as that immunizations are riskier than they are beneficial) and by misleading the public about a specific medical problem.

One of the methods that SM outlets have adopted is programmatic content moderation, although this procedure is quite resource-intensive, may be quite laborious, and may not work consistently across various types of data. There are also strong connections between the signs of technological addiction and mental health issues in relation to SM and addiction.

Additionally, increasingly sophisticated technological uses, like virtual reality, have been linked to detachment and a less "feeling of presence" in objective reality. In order to minimize repetition of information, the reader is directed to past chapters in the chapter that covered various psychological nuances and elements of SM use.

While social media can be a great platform for candid conversation on a wide range of subjects, the unsettling truth is that it may encourage a culture where there are fewer clear-cut distinctions between "real-life" and "social media" friends (i.e., "real-life friend" versus "social media friend").

There is a danger of "invasion" whether it is deliberate or unintentional as a result of the clear demarcation of a specified personal space no longer being. Although SM's drawbacks have been discussed, there is no denying that when used responsibly, SM helps raise public awareness of a variety of health and mental health issues, which can have a positive overall impact.

There are instances of SM working well as a tracking tool for actual disease outbreaks and epidemics. However, the possibility of "false alarms" does occur, which might have an impact on the usefulness of SM as a population-level tool for tracking public health.

Influential personas without the necessary knowledge or subject-matter experience frequently promote the tracked data despite its lack of specificity, potential for distortion, and potential for misinterpretation. It might be difficult to undo the harm caused by misinformation and factual distortion. One dispute that shows how global issues can be misrepresented in a highly visible way is the debate over the claimed link between childhood immunization and autism.

Despite the fact that numerous studies have failed to conclusively establish a link between childhood vaccinations and autism, sizable segments of society continue to support the opposite position.

Unfortunately, instead of authorities and the medical community, mistaken reliance frequently tends to be placed in messengers and various SM tools. These factors need to be taken into account from the standpoint of IHS, particularly when one knows that the "most prominent voice" is not always the one with the best or correct answer.

The threat of malicious actors utilizing social media to distribute false information that can hurt third parties is very real and current. In the middle of the 2009 swine flu season, there was a noticeable increase in SM reporting of numerous theories regarding the flu virus, its purported genetically engineered origins, and other baseless allegations.

With literally thousands of other search results on the subject of the "swine flu epidemic," the implicated SM platforms reported thousands of user views of the dubious content, which was most certainly a huge underestimation. Despite their incorrect assumptions and clear danger, such forms of misinformation are nonetheless prevalent.

For instance, the most recent corona virus outbreak in Wuhan, China, is the subject of numerous erroneous rumors in SM publications. The so-called "social media virality risk," as described by Vijaykumar et al., aims to assess the population's perception of public health risk in the broader context of "social media effectiveness." Risk Amplification through Media Spread (RAMS) is their conceptual framework.

Of paramount relevance, individuals (or groups) who believe they have no other ways to communicate their opinions may favor SM platforms. As a result, SM can be viewed as "virtual aggregators" for those who have opinions that are generally comparable or complementary.

There is also a non-trivial risk that these "virtual communities" will develop beyond their "online presence. "In the past, unconventional or unpopular opinions frequently suffered from social marginalization and lacked access to effective forums for thought and discussion. With social media actively supporting the gathering of like-minded people into organizations, practically every concept can now have a "online home." And while "virtual homes" can serve as centers for innovation, creativity, scientific research, understanding, diversity, and concept exploration, they can also serve as the sources of false information that is harmful to the community. Last but not least, there is at least some evidence that SM may also provide a platform for people to investigate various points of view, thereby opening up opportunities for "fresh viewpoints and open mindedness."

7.9. HUMAN TRAFFICKING

The subject of sexual or other forms of human trafficking involving men, women, children, and smugglers/traffickers was covered in several of the earlier sections. Here, we want to make it clear that human trafficking—which involves people fleeing not only poverty but also violence, abuse of all kinds, and the desire for political asylum out of fear of retaliation—is widespread, virtually unavoidable, and a stain on the struggle for justice as well as on the tools and institutions of government.

"Irregular mobility" is a term that can also be used to describe large-scale migrations of refugees and migrants. There isn't much data on when, why, or how such susceptibility develops in large-scale migrations or refugee movements. The phrase "fluidity" may have caught the reader's attention because there is no clear or linear relationship between the identities of "migrant," "refugee," and "trafficking victim." A person may become a migrant or refugee as a result of trafficking, or migrant or refugee may become a victim of trafficking while fleeing danger. There are various forms of exploitation involved in human trafficking, such as forced marriage, sexual and labor exploitation, trafficking for organ harvesting, trafficking for criminal activities, and trafficking for sexual exploitation and forced labor (Figure 7.4).

Figure 7.4. Human Trafficking.

Source: Image by Flickr

According to Brunovskis and Surtees, those who are trafficked have specific vulnerabilities, including a lack of resources, an absence of legal status, a lack of knowledge about laws, rights, and assistance, language barriers, a lack of access to humanitarian aid, work situations that can result in exploitation and abuse, threats to one's own and one's family's safety, and exposure to violence and abuse in communities and families.

People who move around erratically or who are migrants or refugees camped out for extended periods of time are particularly prone to being trafficked. There is no State institution or tool that can offer adequate protection against exploitation.

A multi-agency approach is necessary to safeguard migrants and refugees, and it should involve regional, national, and international organizations working in "criminal justice, judicial engagement, human rights, and development."

7.10. THEORIES OF MORALITY

Identifying what is right and incorrect behavior is the goal of moral theories. They enable people, and subsequently nations, to critically assess their choices in terms of impact that go beyond social and economic repercussions. The human species has been able to coexist peacefully and to advance the wellbeing of all. There are numerous ethical theories. Among the theories that have received much study are utilitarianism, the Rawls theory of justice, Kantianism, virtue theory, the four-principles approach, and casuistry.

7.10.1. Utilitarianism

According to utilitarianism, morality should strive to enhance the well-being and happiness of humans as a species. According to utilitarian philosopher John Stuart Mill, actions are right in the degree to which they tend to promote happiness and vice versa, where happiness is defined as desired pleasure and the absence of pain.

This moral philosophy is in favor of numerous medical interventions, including social medicine and triage. But it disregards factors like the person's age or our commitments to certain people, like our parents or children.

The amount of self-sacrifice required by this idea makes it challenging to support. If doing so will maximize good or utility, it compels us to sacrifice our interests and the interests of those closest to us for the sake of unknown parties. For instance, it might be challenging to persuade a doctor from the developed world to leave their family behind or put them in risk in order to work in a developing nation. Further, convincing a doctor in a developing nation to forego chances in affluent nations is challenging.

According to utilitarian reasoning, the large quantity of immigrants justifies allocating resources to providing healthcare to this vulnerable population. In other words, compared to the suffering of the numerous refugees, the financial burden is not particularly heavy.

7.10.2. Rawls Theory of Justice

The distinction of Rawls' Theory of Justice is that it takes into account the moral, cultural, and experiential differences that make up our current moral framework. A system or solution for a problem must be created objectively, according to Rawls, who recognizes that social structures skew human perceptions and favor one side of our judgments of morality over the other.

He said that a solution should be developed under the assumption that the person in question has total authority, and that once it is put into place, the person will be returned to society with random characteristics like sex, color, socioeconomic status, or prior oppression or wealth.

For instance, in many nations around the world, students from a less selective American institution are regarded as having more status than those from a top-tier Pakistani university. Even if it may be argued that the education at a third-tier university in the United States is superior to that at a top university from a developing country, this is also not the case as shown by standard test scores.

The best conceivable moral answer to the dilemma could be found using Rawls' theory if it were applied to selecting admission standards. In light of these ambiguities, Rawls contends that an impartial attitude known as "The Original Position" (OP) may be the most effective means of ensuring that lawmakers and politicians, who are in charge of all persons who reside in that area, make judgments that are not impacted by their biases.

The OP aims rationally for systems that have a high likelihood of fostering progress and reducing social unrest while attempting to be objective. In order to reduce biases while making system decisions that affect a large number of individuals, it is assumed that the policymaker will have no control over their features once they are reintroduced into the system. The classic "privilege walk" exercise showed how those with less privilege must put in twice as much effort. Additionally, Rawls' ideas demonstrate how hegemonic institutions, like nations and organizations, build structures to support their supremacy.

According to Rawls, the only way to establish social stability is to elevate everyone to a position of equal moral worth. The distribution of resources from the wealthy to the poor was something he supported, but he also believed that opportunities should only be based on innate abilities and a desire to succeed.

Without addressing immigration directly, Rawls' Law of Peoples adds to the Theory of Justice by offering perception into universal ethics and a concept of justice. He let go of the idea that society was a closed system or a nation-state.

He elaborated on the ideas of noninterference, respect for human rights, and support for nations lacking the prerequisites for the emergence of a just government and proposed that a decent liberal system must be enjoyed by all people.

Rawls is frequently criticized for the practical difficulties in putting a true OP into practice. Furthermore, stringent equality rules are rarely supported because it is difficult to discriminate between the weak, the dishonest, and the underprivileged. In any case, this has been seen as an excellent moral principle to follow while deciding on international health policies and the security of global refugee health.

7.10.3. Moral Subjectivism

According to moral subjectivism, there are no objective moral qualities and that each individual determines what is right and wrong. According to moral subjectivism, no ethical judgments are absolute truths; rather, they reflect the subject's attitude or view.

Moral realism, on the other hand, contends that moral principles are distinct from one's own personal viewpoint. Since all actions have been approved by the topic under consideration, a subjectivist cannot rationally object to anyone's behavior.

Several challenges to the hypothesis have been made. Although the sentence may be unclear, when a subjectivist discusses an ethical dilemma, they are simply conveying their beliefs and not any facts. For instance, a moral subjectivist's claim that "He is a liar" is just a subjective opinion rather than an objective declaration of what they believe to be an immoral conduct.

There are different types of moral subjectivism:

- **Simple subjectivism.** The view that all ethical thoughts are not objective truths but personal feelings and attitudes.
- **Individualist subjectivism.** Protagoras suggested that every human had a distinct moral compass and that his self-interest is the goal.
- **Moral relativism.** Relativism is an extension of moral subjectivism to the society. This view holds that in order to be right, it needs to be approved by the society as well.
- **Ideal observer theory.** This hypothesis was put forth to explain the prejudices and illogical beliefs that people have. If there were too many disagreements, a hypothetical perfect observer would decide.

7.10.4. Theistic Morality

According to this notion, morality is correlated with religious teachings. Theistic voluntarists believe that God's moral laws, or "moral responsibilities," regulate our moral behavior. It is believed that the foundation of our moral obligations is God's command and/or God's will (what he desires for us). According to religious philosophers all people were made in the image of God.

God's instructions serve as a roadmap for carrying out his desire. According to this view, moral decisions can be made about whether a course of action is right or wrong.

Christian Miller clarified the theory by focusing on the following topics: scope (Who is expected to live up to a specific desire of God?), objectivity (the inherent morality of certain acts due to the creation of morality by an all-loving knowledgeable being), learning (from religious texts, leaders, revelations, and reasoning), and nontheist morality (atheists are also created in the image of God and therefore with a properly functioning reasoning can grasp some of the reasons that underlie God's desires.

This hypothesis has been disproved using both "moral constructivism" and "independent moral realism," as well as arguments for and against the existence of a supreme being. The religious texts of the main global faiths, including Judaism, Christianity, Hinduism, and Islam, define it as a moral imperative to care for the ill and weak. Further, theistic morality is violated by healthcare policies that exclude all migrants or a certain group of refugees.

7.10.5. Kantian Ethics

According to Kantian ethics, the intended effect of a given action determines whether or not it is moral to commit that act. According to Kant, it is essential to consider the "maxim," or intended purpose, for our conduct in order to assess its morality.

For instance, in accordance with this notion, doctors are expected to recommend the patient's best course of therapy. It would not be regarded as morally immoral if the prescription had a negative result. Contrarily, if a doctor recommended a medication in order to increase his income, that decision would be morally reprehensible. Only goodwill, according to Kant, can be morally right.

One of the objections is that, according to Kantian morality, a nation's political structure would be responsible for encouraging individual kindness, and that this moral imperative would take precedence over any concerns about utility.

For instance, Kantian morality holds that it would be morally wrong to prevent refugees from receiving healthcare simply because of the act's nature and how it will affect the individuals it is supposed to help. It would disregard any potential negative effects on the nation's citizens and authorized inhabitants.

Kant also fails to propose a hierarchy of moral obligations. For example, can one risk the death of a refugee to prevent the death of a legal resident? Can one lie to prevent the murder of their family? Although Kant offers a moral standard, he also falls short in defining a compelling logical justification for adhering to his beliefs.

7.10.6. Four Principle Theory of Morality

Another framework for critically assessing the morality of medical decisions is provided by justice, autonomy, beneficence, and non-maleficence. Justice indicates that healthcare resources must be allocated in a fair and just manner.

Autonomy allows for individual decisions regarding healthcare. Healthcare professionals have a moral duty to behave in the patient's best interests, which is known as beneficence. The goal of non-maleficence is to keep patients safe.

Although straightforward, this framework is not all-inclusive, and important big-picture judgments need to be judged in accordance with other moral norms.

7.10.7. Arguments Against Providing Healthcare to Refugees

Concerning refugees and, thus, the provision of healthcare services for this population, there is currently no agreement. While some nations have upheld the laws designed to defend their human rights, others have remained silent or made it known that they do not support them.

Many politicians and philosophers have opposed providing medical care, help, food supplies, or shelter in the event of war, famine, or other global disasters. Below is a list of some of the most important claims, followed by an analysis of their veracity and importance.

7.11. HEALTH AND SAFETY AT WORK

This section will not examine the many measures used to promote employee health and safety at work or the various ergonomic and safety standards developed for the workplace. To understand the nuances of actions related to health and safety at work, a thorough discussion of the subject will be conducted.

As a result, a few acts related to health and safety at work will be discussed, but solely with the aim of understanding their nuances. Instead of the prescription that is established in health and safety at work, the description will be the main focus.

Safety and occupational health are related concepts that have expanded our understanding of illnesses and accidents in the workplace. Risks associated with the workplace are now considered from a variety of viewpoints, including biological, ergonomic, and psychological ones, in addition to the traditional physical and chemical ones.

Additionally, illnesses are no longer just considered to be occupational illnesses brought on by the workplace; they are now also considered to be job-related diseases and illnesses that are made worse by employment.

Changes in the modes of production and management of many businesses and organizations went hand in hand with these advancements in our understanding of illnesses and accidents. The risk associated with work have evolved from the Taylorism-Fordist-managed industry to flexible manufacturing (Toyotas' model).

In view of this, efforts to quantify and categorize the risks of illnesses, fatalities, and accidents in the workplace have been made using both technological and managerial criteria. Risk assessments were used to make diagnoses of the work and organizational settings, which led to the development of strategies to reduce and control risks.To a considerable extent, quantitative methodologies are applied for workplace hygiene initiatives and qualitative techniques for workplace safety actions. Despite the fact that the many activities for health and safety at work have benefits for people, the emphasis is still on technical rather than human qualities. This emphasis causes a gap between specified labor and actual work since people naturally adapt their work to the circumstances under which it is performed.

The technical aspects of managing health and safety at work also reflect this discrepancy between prescribed work and actual work. As a result of the complexity of the developed operations and the working environment, the containment of one risk can lead to the development of new dangers.

For instance, the use of personal protective equipment may cause the worker to develop some sort of allergy, precisely because the technical aspect of reducing a certain risk was considered, but not the adaptability of the material to the human body.

Furthermore, what might be considered the minimal tolerable amount of risk at work may not immediately cause a problem but may eventually lead to illnesses, accidents, or even fatalities. For instance, exposure to a specific chemical element might not be dangerous in isolated exposures, but it is a dangerous chemical element precisely because it has accumulated in the human body.

Workplace health and safety initiatives always consider factors pertaining to the majority and invalidate the particular elements of employees. Instead of the other way around, the human person is still being adapted to the workplace.

As such, health and safety at work initiatives frequently place a greater emphasis on the hazards posed by physical, chemical, biological, and ergonomic factors than by psychosocial factors. This is due to the fact that these psychosocial risks are associated with personal characteristics that do not often concentrate cases.

With this, the issues created in health and safety at work are highlighted by both the disparity between prescribed work and actual labor as well as the issues of these acts' unequal treatment. As a result, some equally significant reflections are prompted, which does not lessen the significance of the activities or their existence in workplace health and safety. If incidents, fatalities, and diseases attributable to the workplace continue to occur, something about the policies and procedures intended to promote health and safety at work has to be addressed. There is a degree of non-compliance with these required measures that is not evident when considering the number of workplace accidents, fatalities, and illnesses.

These disparities and this degree of noncompliance are related to a bigger issue, which is how people live their lives. Although it may seem that the main concern is with life, there are other aspects of work in organizations and sectors that better account for investments in health and safety at work as well as disparities in the measures taken.

Investments in workplace health and safety strive to minimize or cut down on the costs of accidents, diseases, and deaths brought on by the workplace. Because of this, the emphasis has traditionally been on hazards that arise in the workplace, ignoring conditions that are made worse by the job. Since the complete eradication of risk is somewhat utopian, they demonstrate a minimum degree of tolerance to physical, chemical, biological, ergonomic, and psychosocial dangers, similar to many other acts in workplace health and safety.

The importance of these people's lives, however, is at the heart of this debate. What level in the organization or industry are they at in terms of hierarchy, the risks they face, whether or not their role is crucial to the organization or industry, whether or not costs associated with illnesses, accidents, and fatalities are high, and whether or not breaking the law is preferable to operating legally? (not for the individual but for the industry or organization). The analysis of disparities in workplace health and safety actions will be covered in the following section of this chapter. The political level of life, which is crucial to workplace health and safety, is something that both Foucault and Agamben consider.

7.12. ADVANTAGES OF MICROFLUIDIC LAB-ON-A-CHIP TECHNOLOGY

Every nation worldwide has similar difficulties in maintaining the wellbeing of its citizens and preventing the spread of contagious diseases across borders. In the context of ensuring the security of global public health, biomedical engineers, synthetic chemists, biologists, and public health experts are assessing the possibilities of microfluidic LoC technology.

The detection of infectious diseases and diagnostics for global public health are only a few of the applications that have evolved in recent years.

Cell analysis, cytometry, blood analysis, nucleic acid amplification, genetic mapping, enzymatic tests, peptide analysis, protein separation, toxicity analysis, and bioassays are some more on-chip clinical evaluations that have been developed.

The importance of cell targeting, clinical trials, drug production, pharmaceutical formulations, and the process of product management has led to a gradual increase in the importance of LoC devices in the field of drug research. The LoC devices are considered promising for medication analysis and dosage optimization.

This is extremely helpful for evaluating the synergistic effects of medications taken in combination. Microfluidic LoCs recently offered an exclusive opportunity to simulate natural veins for evaluating nanoparticles as drug carriers for targeting cells or, furthermore, provided possibilities to examine the in vitro metabolism of biological cultures.

Technical advantages of LoC devices include their portability, compactness, modularity, embedded computing, automated sample processing, low electronic noise, limited power consumption, and ease of component integration. Additionally, LoC devices can handle a variety of operations such sampling, routing, transport, dispensing, and mixing, typically with fewer moving or rotating mechanisms.

The LoC devices' compact size allows them to provide precise fluidic conveyance using electro kinetics or micro pumping, effective liquid sample separation, and accurate sample measurement. Similar to this, the LoC devices' ability to manage small fluidic volumes allows them to speed up processes like product manufacturing and sample processing. As a result of its capacity to efficiently manage the concentration of the reactants, LoC devices also enable more accurate measurement of samples.

The same is true for clinical medicine, which has greatly profited from microfluidic LoC technology because it is suitable for a variety of tests including disease detection, pandemic monitoring, glucose monitoring, diabetes control, and many other procedures.

The use of LoC devices improves a variety of biological tests that need the mixing, analysis, and separation of samples, which often include cell suspensions, nucleic acids, and proteins; analytical, electrical, or optical detection methods are also feasible.

The following key manufacturing benefits make microfluidic LoC technology practical: attainable mass production, low replacement cost, quick manufacture, straightforward quality checks, and wide range of supporting computer-aided design and simulation software tools. However, there is still room for improvement in the design of microfluidic LoC technology due to technical issues such size reduction, sample input rates, power consumption, chip durability, and biocompatibility.

7.13. CONCLUSION

This chapter concludes about the Contemporary developments and perspectives in International Health Security. The scope of health security, Different emerging infectious diseases like COVID-19, Ebola, etc. Chronic health conditions and its care, Trauma system, Emergency medical services.

Ransoware types and payment have been discussed in this chapter. Detailed understanding of International Health Security. Human Trafficking justice for human, exposure to violence and abuse, sexual and labor exploitation. Theories of mortality and morbidity. Health and safety and different health challenges addressed in the tourism sector.

REFERENCES

1. Baig, M. A. A., (2021). mHealth-based microfluidic lab-on-a-chip for international health security. *Contemporary Developments and Perspectives in International Health Security* (Vol. 1). [Preprint]. Available at: https://doi.org/10.5772/intechopen.90283.
2. Conti, K., et al., (2021). The evolving interplay between social media and international health security: A point of view. *Contemporary Developments and Perspectives in International Health Security* (Vol. 1). [Preprint]. Available at: https://doi.org/10.5772/intechopen.93215.
3. Le, N. K., et al., (2021). International health security: A summative assessment by ACAIM consensus group. *Contemporary Developments and Perspectives in International Health Security* (Vol. 1) [Preprint]. Available at: https://doi.org/10.5772/intechopen.93214.
4. Luciano, M., (2021). Beyond health and safety at work: Reflections on biopolitics in occupational health as an important component of international heasslth security. *Contemporary Developments and Perspectives in International Health Security* (Vol. 1). [Preprint]. Available at: https://doi.org/10.5772/intechopen.93749.
5. Miller, A. C., Khan, A. M., & Ziad, S., (2021). Ransomware and academic international medicine. *Contemporary Developments and Perspectives in International Health Security* (Vol. 1). [Preprint]. Available at: https://doi.org/10.5772/intechopen.91762.
6. Papadimos, T., et al., (2021). Health security and the refugee crisis in greece: The refugee perspective. *Contemporary Developments and Perspectives in International Health Security* (Vol. 1). [Preprint]. Available at: https://doi.org/10.5772/intechopen.91210.
7. Reis, N., & Cipolla, J., (2021). The impact of systems of care on international health security. *Contemporary Developments and Perspectives in International Health Security* (Vol. 1). [Preprint]. Available at: https://doi.org/10.5772/intechopen.93055.
8. Sparkle, T., & Roy, D., (2021). A moral perspective on refugee healthcare. *Contemporary Developments and Perspectives in International Health Security* (Vol. 1). [Preprint]. Available at: https://doi.org/10.5772/intechopen.92110.
9. Stawicki, S. P., et al., (2021). Introductory chapter: International health security expanded and re-defined. *Contemporary Developments and Perspectives in International Health Security* (Vol. 1). [Preprint]. Available at: https://doi.org/10.5772/intechopen.94394.

Chapter

8 TOURISM: FROM EMPIRICAL RESEARCH TO PRACTICAL APPLICATION

Contents

8.1. Community Participation Toward Tourism Development 202

8.2. Types of Community Participation 205

8.3. Factors Influencing Community Participation 206

8.4. Community Development Strategies for Tourism Development in Langkawi Islands, Malaysia 207

8.5. Ecotourism and its Role in Sustainable Development 211

8.6. Tourism, Competitiveness, and Economic Growth: A New Analytical Model 218

8.7. Interaction Between Cultural/ Creative Tourism and Tourism/Cultural Heritage Industries 223

8.8. Conclusion 229

In the chapter of tourism from empirical research to practical application it talks about the community participation toward tourism development. This chapter also explains the types of community participation, factor influencing community participation. This chapter also addresses the ecotourism and its role in sustainable development.

The chapter also addresses the tourism, competitiveness and economic growth such as measurement indicators, conceptual framework of tourism competitiveness, economic growth, and tourism activity. At the end of the chapter, this chapter explains the interaction between cultural/creative tourism and tourism/cultural heritage industries.

8.1. COMMUNITY PARTICIPATION TOWARD TOURISM DEVELOPMENT

A community is defined as a collection of individuals who reside in the same region and are connected by shared ethnicity, culture, values, or socioeconomic status. A "community" in a World Heritage Site (WHS) location refers to the locals who play a crucial role in preserving the WHS. As a result of their cooperation in attaining shared objectives and improving the community as a place to live, community members form a relationship that is defined as involvement in the community.

Conflicts between citizens' demands and interests, the pursuit of economic development and a higher standard of living, and WHS conservation can be resolved by community involvement in heritage management.

According to the Charter for the Conservation of Historic Towns and Urban Areas, "The participation and the involvement of the citizens are crucial for the success of the conservation program and should be promoted." . The Burra Charter also emphasizes that community engagement is essential for heritage conservation to be sustainable.

Several studies have concentrated on community involvement in WHS conservation and the growth of heritage tourism. It is crucial for WHS locations to be revitalized and to maintain their development throughout time that the community participates in WHS conservation and tourism development.

Local people's involvement in WHS heritage management and tourism development benefits to both their quality of life and the long-term preservation of the heritage site. Further, locals who participate feel more a part of the community, build stronger social networks, and acquire a deeper respect and understanding of the importance of the neighborhood. Coercive participation, induced participation, and spontaneous participation are the three categories of community participation that can be distinguished in the context of tourism development and heritage management. The lowest level of community involvement, in which citizens have no control over how the tourism industry develops, is referred to as coercive community participation.

Their involvement is restricted to a variety of predetermined actions focused on promoting tourism destinations, and they don't reap many financial benefits. Locals participate in induced community participation, but while they have a voice in the heritage management and tourism development process, they have little real influence or control over the decisions that people in positions of authority make. Locals have decision-making authority and control over the development process through spontaneous engagement.

Nevertheless, the contribution of local people to WHS conservation and tourism development depends on a number of variables that may either help or impede their contribution. Local populations must be aware of the importance of the WHS in which they live and possess the required knowledge and abilities to take advantage of tourism development and conservation opportunities.

However, in order to encourage locals to participate in tourism development and conservation efforts and prevent these chances from being lost, some incentives may be required. Residents may be inspired to participate in tourism development and WHS conservation if they perceive benefits from the designation of a site as a WHS and from the development of the tourism industry.

8.1.1. Participation in WHS Conservation and Tourism Development

As was previously mentioned, community participation is a relationship that is created by the community members themselves via their mutual cooperation in reaching shared objectives and improving their community's quality of life. Therefore, community involvement is the practice of collaborating with members of the community for the benefit of that community.

In such a setting, community members' interactions and connections play a crucial role in forging lasting friendships. Therefore, becoming active in the community can provide residents a sense of identity, credibility, and trust. Numerous studies support the significance of community involvement in the preservation and growth of tourism at heritage sites. Community involvement in WHS management can help resolve tensions between the community's economic and developmental goals and the need to preserve the WHS destination as a priceless resource. It can also help community members better understand what heritage is.

The significance of public involvement in sustainable conservation programs has been supported by numerous heritage management studies. Participation of the local community in heritage management promotes economic growth, improved quality of life, and the sustainability of conservation initiatives. Participation in community activities can also instill a sense of pride in the community.

The UNESCO World Heritage Committee asserts that the designation of a location as a World Heritage Site (WHS) is a component of a comprehensive plan to recognize and safeguard the location. Therefore, it is ideal for the community to live in harmony with the value linked with the site as a result of the acquired local knowledge and sense of belonging.

To familiarize themselves with their legacy and experience a sense of pride for their place and historical treasures, however, local populations must first be educated about their histories. This process of creating pride in the community depends on community participation (Figure 8.1).

Figure 8.1. 40th Edition of UNESCO World Heritage Session.

Source: Image by Flickr

Participation in heritage initiatives by the community can enhance citizens' sense of pride in their community and their awareness of its importance, as well as their sense of belonging and ability to form social networks.

Although there are tensions and symbioses between WHSs and tourism growth, they nonetheless exist. The designation of a location as a World Heritage Site raises its profile internationally and encourages tourism growth, which may result in increased public and financial support for the preservation of the designated area.

However, it is crucial to promote the location as a tourist attraction in order to boost the local economy. At the same time, sustainable conservation programs must be given top priority in order to preserve the location itself. Therefore, community involvement in WHS conservation programs and tourism growth is important and valuable.

Because it entails fostering relationships between community members, which in turn contribute to the development of a sense of legitimacy, trust, and belonging, community engagement improves communities.

Additionally, community involvement in WHS conservation and tourism development is an empowering process that involves all levels of stakeholders (local government, local residents, and private enterprise), so that problem identification and decision-making are shared and stakeholders have a shared interest in the sustainability of the development. Their participation in this process benefits both the local community and the tourism developers, increasing their appreciation for the locals' traditional way of life and values.

Destination communities should be actively involved in tourism planning because they have a historical understanding of how the community adapts to change and are the group most impacted by tourism. This is especially true given the expectation that they will become a crucial component of the tourism product. Moreover, by involving the community in the planning process, some of the negative effects of tourism may be minimized and the beneficial effects may be amplified.

8.2. TYPES OF COMMUNITY PARTICIPATION

Many academics have recognized different forms of community involvement, from manipulative participation to citizen power. Arnstein, a pioneer in this field, proposed a hierarchy of eight levels of resident engagement divided into three categories: manipulative participation, citizen tokenism, and citizen power.

Pretty created a typology of community involvement with three types that includes elements of manipulative participation, passive participation, and self-mobilization. Tosun used Arnstein and Pretty's typologies to analyze tourism and define the three types of community participation already discussed: coerced participation, induced participation, and spontaneous participation.

Notably, however, residents' involvement in coerced community participation is confined to preset activities in accordance with decisions made by power holders who also decide how people will act to promote the destination and to what financial rewards they would be entitled. However, the locals themselves have no real influence or opportunities to have their concerns heard.

According to Tosun's typology, induced community engagement is the second type of community involvement. It is comparable to citizen tokenism in Arnstein's model and consultation in Petty's typology. Induced community engagement gives locals a voice in the tourism industry but denies them the ability to influence how decisions are made.

8.3. FACTORS INFLUENCING COMMUNITY PARTICIPATION

The choices of residents for community involvement depend on a number of variables. Having looked at the variables affecting locals' choices for community involvement in heritage preservation and tourism growth, three groups of influencing variables become apparent: motivation, opportunity, and ability. Motivation refers to residents' interest and willingness to participate in the development/management process.

In the context of WHS locations, resident attitudes, WHS inscription, and future tourism development benefits are influenced by motivation. Focusing on the community's perceptions of the positive effects of tourism increases their support for tourism development and encourages them to engage in tourism-related activities and heritage preservation initiatives. In contrast, focusing on the negative perceptions of the impacts decreases their support for tourism development.

As a result, community involvement in WHS conservation programs and tourism development procedures is influenced by locals' concerns, interests, and perceptions of the effects of tourism development. Keeping this in mind, the designation of a place as a WHS and the growth of tourism can have a positive impact on the lives of the people who live there by increasing their income, providing them with more employment opportunities, raising their standards of living, improving their public infrastructure, expanding their access to recreational and entertainment opportunities, and fostering the preservation of their culture.

However, tourism also has the potential to have a detrimental impact on local communities by boosting the cost of living, increasing the price of real estate, causing overcrowding and traffic congestion, and growing the prevalence of crime and drug use.

Numerous studies have emphasized the importance of perceptions of WHS inscription and tourism growth on community involvement among people of the site. These studies indicate that locals who believe that the benefits of WHS inscription and tourism development exceed the drawbacks are more inclined to engage in WHS conservation initiatives and tourism growth.

However, for the majority of residents of rural destinations, this interest in participation is restricted to taking part in activities that have been predetermined, with decision-makers empowered to act on behalf of the community with regard to how the destination will be promoted and what economic opportunities will be available for the local community.

According to these studies, people who live in WHS destination communities typically don't want to be involved in decision-making.

8.4. COMMUNITY DEVELOPMENT STRATEGIES FOR TOURISM DEVELOPMENT IN LANGKAWI ISLANDS, MALAYSIA

When the local community views tourism as a local resource, there has been research showing a strong and favorable relationship between tourism and economic growth. According to, tourism has been highlighted as a key component in many modern island economies. However, in most developing nations, it has turned into a double-edged sword.

In terms of economic development, tourism offers a way to improve rural lifestyles and bring about improvements in the distribution of income in disadvantaged areas. Local communities will have a substantial amount of control over the activities taking place and receive a big percentage of the tangible benefits if local involvement in the tourism sector is increased. However, the rapid growth of tourism has a lot of unintended negative effects. Locals have seen the benefits of tourism as being extremely limited, while the drawbacks include significant environmental harm, cultural eroding, and communal tensions.

For instance, the overestimated statistics on Langkawi Island in Malaysia has resulted in an excess of hotel rooms and chalets, which has led to wasteful competition among resort operators and the eventual abandonment of some, leaving unsightly scars on the environment.

These problems result from a lack of public participation in the planning process (without taking into account the design) in Langkawi, as well as the restricted participation of the private sector and local community in tourism planning.

When looking at the laws and policies that the Federal Government of Malaysia implemented, for instance, during the 10th Malaysia Plan period (2011–2015), one of the major goals was the empowerment of the local community, particularly women, to enable them to fully utilize their potential to participate in economic and social activities.

The Federal Government and the Langkawi Development Authority (LADA) also developed the Langkawi Tourism Blueprint 2011–2015 in accordance with the Malaysia Five Year Plan, with the goal of transforming Langkawi Island into one of the top 10 island travel destinations on the planet. 'Products,' 'Infrastructure,' and 'Enablers' were the three topics in the blueprint, while 'Capacity Building for Local Communities' was the last theme (Figure 8.2).

Figure 8.2. The Langkawi Development Authority (LADA).

Source: Image by Wikimedia Commons

For instance, one of the initiatives was to open the Langkawi Tourism Academy in order to increase the number of graduates entering the tourism sector and address the talent deficit in Langkawi. Community capacity building initiatives may be an excellent location for the Langkawi tourism sector to start in order to emphasize local participation in tourism development because they enable more community involvement in the management and control of their resources.

The capacity building strategy can help undeveloped communities become more able to engage in tourism decision-making processes, which will help lessen the monopoly of power in local tourism enterprises. This strategy may ultimately encourage local populations to embrace sustainable tourism.

8.4.1. Community Capacity Building in Tourism Development

In the context of tourist development, developing local community capacity is viewed as a "win-win approach" that refers to "community" effort, time, resources, leadership, and dedication directed toward "community." This strategy is the essence of "community development," which aids the local community in enhancing their capacity to take part in tourism decision-making, grow their influence, as well as.

Individual, organizational, and community levels all contribute to the development of a community, and during the empowerment process, all levels can converge to boost underutilized skills. Each level will eventually be able to handle their own affairs and pursue their own development interests. The capacity of all people, groups, and communities to adapt to change, take into account tourism development, and cooperate can encourage and sustain efforts for enhancing the welfare and benefits of the community.

The first step in community capacity building at the micro level is to focus on the person component. This group is the focus of programs for education, skills development, job training, and social well-being. At the organizational level, "community capacity building" refers to the assets, know-how, and procedures connected to local laborers, technology, management plans, networks, and financial resources aimed at enhancing performance and achieving long-term objectives.

Macro-level indicators of natural resources, people, socio-cultural variables, budgets, policies, the political system, education, and societal welfare in tourism operations are all included in the term "community-based empowerment." Community leadership, ecotourism collaborations, community-based entrepreneurship (CBE), training and education, and external support are just a few of the resources and tactics that have been made available to help communities increase their potential.

The importance of community leadership, partnerships formed via cooperation between government agencies, non-governmental organizations (NGOs), the private sector, and locals has been highlighted by community capacity building. When the leaders are able to assure the active participation of a varied network of community members, they can use leadership as a tactic to improve capacity. This enables persons with divergent interests to take collective action by forming a unit of solution.

Community-based ecotourism partnerships are one of the most effective methods for capacity building. In fact, a community's desired form of tourism can be determined by a multi-stakeholder cooperation at all levels without having detrimental effects. Three ecotourism collaborations have been proposed; they can be between a community and an NGO, a community and a private enterprise, or a federation and a private company.

As a result, a genuine partnership may be a useful strategy for completely transferring ownership to local people and achieving reciprocal material benefits through the pooling of resources and expertise to promote ecotourism.

CBE is likewise regarded as one of the community-beneficial strategies that has been successful. This strategy was established to empower the local community as small business became a crucial aspect of the tourism sector worldwide. It is essential for both short-term and long-term benefits that local entrepreneurs from new businesses operating in the neighborhood get involved. To further inform locals about the effects on their community, education and training programs are required in the tourism sector.

According to this approach, the community's understanding of tourist growth will grow, the groundwork for local tourism leadership will be laid, and most crucially, tourism knowledge awareness will be generated. The optimal time to provide education and training programs is in the beginning stages of building community capacity because they promote tourist knowledge and awareness. For instance, certain training programs might expose participants to some of the worries a traveler could have.

The supply of financial resources, technical assistance, training, information, political negotiating, and facilitation of capacity building procedures are all examples of how the outside world can promote the development of community capacity.

The strategy places a focus on the exchange of knowledge, small private and community investors' assistance with product innovation, assistance with market access among the local community, the creation of spaces and social networks for local people to communicate, the provision of information on markets, marketing, and technology, and assistance to local authorities in creating support systems for small tourism-related businesses in the destination.

Consequently, Langkawi communities must implement a capacity building strategy since it gives them the authority to manage resources including real estate, finances, skills, natural resources, and tourism-related knowledge. As a result of this strategy, they may be inspired to acknowledge their abilities, principles, and local knowledge that will enable them to support sustainable tourism.

8.5. ECOTOURISM AND ITS ROLE IN SUSTAINABLE DEVELOPMENT

Ecotourism is one of the fastest expanding subsectors of the sustainable tourism sector, which prioritizes economic development, poverty reduction, environmental protection, and wildlife conservation. The word "Ecotourism," which Hector Ceballos Lascurain first used to characterize nature-based travel to largely untamed areas with a focus on teaching, was first used in 1983.

There isn't a single, agreed definition of ecotourism. It differs from nature-based tourism in that it promotes sustainable rural development and makes biodiversity protection profitable for nearby populations. In many rural communities, ecotourism is one of the most popular methods for community and environmental development. Its ability to protect the ecological integrity of local resources by utilizing them sparingly and without consumption determines how effective it will be at generating local economic advantages.

Ecotourism is a different type of tourism that promotes travel within the biophysical context of natural regions. It combines environmentally sound practices, conservation-supporting policies, and community participation. It neoliberalizes nature and emphasizes economic growth, neighborhood improvement, the eradication of poverty, the protection of wildlife, and the environment. Traditional economic pillars including farming, raising livestock, and hunting are incompatible with protected areas; therefore, traditional economic activities can be changed through influencing cultural attitudes in favor of promoting ecotourism (Figure 8.3).

Figure 8.3. Ecotourism: Tourists waiting to take photographs of orangutans.

Source: Image by Flickr

Employment opportunities and a market for regional goods are provided by ecotourism. In addition to promoting cultural awareness in host-guest interactions, it serves as a driver for environmentally sustainable growth. But in order to handle the concerns of ownership, management, and coordination of protected areas, planning is necessary. It improves the protection of wildlife and the fair distribution of ecotourism profits. It promotes local community well-being and aids in the preservation of built, cultural, and natural resources.

It balances the competing goals of economic development and biodiversity conservation while having the goal of environmental conservation. Sustainable economic development is also achieved. It achieves the objectives of preserving biodiversity, reducing poverty, and ensuring corporate sustainability when using sustainable ideas and methods. Locals appreciate the value of environmental preservation and take active steps to safeguard it through ecotourism. They uphold national requirements for all types of facilities, including drinking water, sewage, lampblack, and the atmosphere and sound quality.

8.5.1. Different Forms of Ecotourism

The best way to make money from running a tourism business in an area that is protected, owned, and managed by a community is through community-based ecotourism. Environmental protection, business enterprise, and community development will all have direct and indirect participants and beneficiaries. Tourism development in such locations requires local employment, biodiversity preservation, and cultural preservation. To successfully execute such ecotourism, accessibility must be improved, pricing policy must be liberalized, the community must be empowered, tourism facilities must be developed, and nature-based tourism products must be incorporated. Community-based ecotourism is connected with modest lodging and services promoted for independent and low-budget travelers who do not anticipate high-end tourist amenities. Although there is a high level of public participation, it occasionally excludes community interests and favors a small elite. In this type of tourism, wildlife and landscapes are marketed in a variety of ways as pictures, products, and destinations. Ecosystem tourism is a supply-driven tourist strategy that promotes sustainable development with the ecosystem integrity as its main focus. It does not prohibit exploitation of the environment by tourists and businesses, but the management of entire ecosystems or a particular type of biological variety is based on how it affects the tourism industry.

Figure 8.4. Xiaodubai Rural Tourism Area scene.

Source: Image by Wikimedia Commons

The ecosystem, which is made up of assemblages of creatures, the physical environment, and a variety of interactions and feedbacks, serves as the primary resource base or capital stock for tourism. In this type of ecotourism, the environment is able to withstand or adjust to the stresses of visitors until it takes on a more metropolitan appearance (Figure 8.4).

The term "rural tourism" refers to a type of tourism that developed in Europe and is based on rural resources that depict rurality, traditional folk customs, and agricultural products. In this type of tourism, visitors take pleasure in lounging in the yard, strolling and admiring the flowers, greenery, and gardens, as well as playing cards and looking at the historic buildings.

By fusing agriculture with tourism, it boosts agricultural revenue and gives peasants greater earnings, more employment options, and better living conditions. It stimulates the growth of rural culture and economics in a beneficial way. Rural tourism is in its infancy, with lodgings and basic entertainment taking a back seat to agricultural sightseeing, farming experiences, and close-quarters leisure.

The preservation of cultural traditions through cultural tourism improves the standard of living in the areas it serves. To develop and promote minority cultures and cultural traditions, it is essential to offer education and training programs on ecotourism and nature reserves to tourism business owners, staff members, and visitors. To attract scientists and students to conduct research in the area, chances for research should be made available.

A significant and expanding global industry, nature-based tourism is dependent on the natural surroundings found in parks and other protected places. It is based on sufficient levels of consumer service and suitable levels of environmental quality. In many nations, nature-based tourism has played a significant role in the tourism sector.

8.5.2. Ecotourism in Global and National Scenario

International tourism in developing nations is heavily based on protected areas, particularly in parks and other protected regions. It has enhanced local livelihoods that are directly dependent on tourism as well as brought in a sizable amount of money for the state. The establishment of protected areas and ensuring their long-term sustainability are supported by numerous state governments and international conservation organizations.

Kenya, Zimbabwe, Tanzania, Costa Rica, and Nepal are just a few examples of nations where protected area tourism accounts for a sizable share of international travel. There is nature-based tourism in Australia, Tanzania and New Zealand. The national ecotourism strategy for Australia had focused on ecotourism in regional areas to generate foreign exchange earnings, employment, and other economic and social benefits. For the aggressive and successful policy development in the country, Australia is taking advantage with its spectacular and diverse natural features, unique flora and fauna, and diverse cultural heritage.

After the coronation of 1974 in Bhutan, small groups of tourists were allowed into the country and given permission to visit dzongs and goembas in Thimphu and Paro. Government established a quota of 200 tourists a year in which tourist have to travel in a group of six or more. The cost was set at US$130 per day and all the guests had to stay in government-approved hotels, guesthouses and lodges. Bhutan's rich heritage and vibrant culture has its distinctive traditions and customs demonstrated in its religious festivals and architecture.

Towering crags and highly diverse forests is also home to takin, snow leopard, golden langur, red panda, tiger and elephant. Trained trekking guides are provided periodic knowledge and services to accompany trekking groups. Gross expenditure of US$14 million provides significant foreign exchange earnings in Bhutan.

The current approach entails a fixed price established by the government, with government regulation, tour operator market share competition, and quality of service maintenance. Bhutan's ecotourism has helped Western nations reduce their negative cultural and environmental effects there.

Local communities participate in ecotourism initiatives in the Small Islands of the South Dodecanese and work to reduce conflict. The islands' geomorphology and location offer opportunities for sailing, mountain climbing, and nature hiking. It helps in extending the tourist season and boosting tourist spending. Adventure tourism activities like free-diving underwater fishing have very little negative ecological effects.

In nations like Costa Rica and Guatemala, ecotourism is already the main source of foreign currency. It is one of the important drivers of economic growth and employment creation. Evidence from around the nation shows that tourism is labor-intensive and provides a variety of small-scale opportunities that create jobs for underprivileged, female, youthful, and indigenous groups.

In the southern region of Thailand, community-based ecotourism has mixed results. In terms of fostering community growth, the advantages of ecotourism in Phuket outweigh the expenses. The political, economic, social, and environmental interests of host communities can increase as a result of ecotourism's promotion of environmental sensitivity and responsibility.

Community-based ecotourism and empowerment have a complicated and unfinished relationship. By empowering communities politically and socially, individuals can achieve economic and psychological independence. In Thailand's rural areas, there were also many nature-based tourism businesses.

8.5.3. Ecotourism and Its Impacts on Environmental Aspects

8.5.3.1. Environmental Impacts: Positive impacts

Ecotourism is a supply-driven approach to travel that promotes the integrity of the ecosystem as the key to sustainable growth. It is determined by the ecosystem's capacity to withstand influences but does not forbid commercial and tourism exploitation of the ecosystem. Ecotourism merely regulates the type and volume of visitor activities; it does not impose visitor numbers restrictions based on the idea of carrying capacity. All enterprises must adhere to the relevant national criteria for environmental conservation with regard to air quality, noise level, drinking water, sewage, lamp black, and other factors.

It has a great deal of potential to produce direct environmental advantages for the community. It helps preserve the ecosystem, preserve rare animals, and attract tourists to generate income. Through a program that gives local committees benefits from tourism, Parks assists in fusing conservation and development. The location of visitor amenities and companies is encouraged by commercial hiking and trekking.

Locals value environmental preservation and take active measures to protect it. Through nature exploration and cooperation between managers of protected areas and tourism businesses, conservation of nature draws tourists and offers high levels of satisfaction. The rate of forest cover can be enhanced, providing a strong foundation for the growth of ecotourism. When developing tourism, ecotourism placed a strong emphasis on preserving the environment and biological variety. It also calls for preserving the sustainability of resource use and turning the benefits of local tourism resources into industrial and financial gains.

8.5.3.2. Environmental Impacts: Negative Impacts

An industrial activity like tourism has a number of effects that are typical of most other industrial activities. It requires specialized infrastructure and superstructure to support it, uses limited resources, and generates waste by-products. Negative effects on the resources of protected areas result from ineffective tourism management. It offers a different method of utilizing resources that could have a negative impact on environmental resources.

Litter problems had occurred in 44% of the nature reserves, water pollution in 12%, noise pollution in 11% and air pollution in 3% of the nature reserves in China. Camping sites felt environmental degradation in both wet and dry seasons but more in dry season. Environmental degradation is a common problem which increases with increase in visitor number in a protected area open to recreational use.

There will be more pollution in the neighboring river near camping and caravanning areas. Local pollution could have devastating repercussions, including fish kills, and the ecology could become less stable. Increased clearing of land and additional visitor space are necessary for ecotourism. A significant decline in forest cover, vegetation destruction, desertification, and soil erosion are some undesirable outcomes of farmers' intensive farming practices.

Successful ecotourism programs may generate more interest and tourists as a result, amplifying unfavorable effects including increased solid waste output, ecological disruption, and route erosion. Such effects may gravely jeopardize the resources that support ecotourism.

8.5.3.3. Social Impacts: Positive Impacts

Despite being very small, tourism has a favorable social impact. People migrated to the area as a result of increased tourism. Guest houses are situated along the primary access route to the National Park and offer food and refreshments. In order to overcome political opposition to economically unsound and environmentally damaging tourism initiatives, economic benefits of tourism must be realized.

Despite modernization, the locals still practice their honest folkways and provide their customary hospitality. As a result of economic growth, the traditional culture has been enhanced. To ensure the experience of both the tourists and the community residents, ecotourism providers may conduct cultural excursions with various community organizations. Visitors can be made aware of the great cultural diversity of the area. Lodges should act as good neighbors to both locals and tourists for sustainable ecotourism. Tourism promotes social harmony and stability.

Ecotourism provides a range of goods and services that fall under the experience and service categories. The main reason people visit ecotourism destinations like historical homes or theme parks is for their emotional or experiential aspects; this is more an example of how people utilize services than it is a reference to the practical aspects of the destinations. Thus, the psychological advantages that Eco tourists experience are seen as more pertinent and significant in this setting. Environmental preservation and plant development do not contradict. The growth of rural tourism gives the people an additional source of income.

8.5.3.4. Social Impacts: Negative Impacts

A tourism-related initiative could encourage social unrest and undermine cultural norms. Conflicts between the populace and the government could arise. As a result of tourism, there is a rise in drug use, prostitution, and alcoholism. Tourists' exposure to foreign cultures affects the traditional and undeveloped cultures of the local area.

Traditional locals with simple lifestyles and material wants are susceptible to Western influence as a result of tourism. Some places find it unbearable when young people dress like Westerners and wear vests. Alcohol usage and gender-based behavior among younger generations go against regional norms. Participation in the idea and execution of their collective ecotourism program has had mixed results. The knowledge and enjoyment of financial progress among community members differs.

8.6. TOURISM, COMPETITIVENESS, AND ECONOMIC GROWTH: A NEW ANALYTICAL MODEL

The study of tourism competitiveness has been influenced by the appeal or factors that influence people's travel decisions. It is possible to draw attention to the models in the theoretical investigations. It is demonstrated that there is no single factor that determines tourism attraction but rather that tourism competitiveness is related to the sum of several factors.

As a result, a variety of indicators have been created that seek to summarize the destination's level of competitiveness. These indicators, which are discovered using different approaches, basically contain a number of elements. They include those pertaining to the provision of tangible resources or those left over from the past.

These investigations are interested in learning more about the variables that affect the expansion of tourism. Other economists have focused their attention on the empirical investigation of the connection between tourism and growth at the same time as these studies. These analyzes first aimed to provide empirical support for the so-called tourism-led growth (TLG) theory, as it was the first study to look at the causal links between the two variables.

Other, more recent studies focus on understanding the mechanism, the manner, and the magnitude by which tourism generates economic growth. They empirically analyze this relationship, taking into account variables related to tourism demand in the production functions, since it is crucial to take other productive factors into account when analyzing the link between tourism activity and economic growth.

8.6.1. Tourism Competitiveness: Measurement Indicators

Since the last decade of the 20th century, an increasing number of researchers have concentrated on understanding the factors that influence a geographical enclave to become a preferred tourist destination over others and the factors that determine their potential for tourist attraction, whether it be greater or less. The ideas have concentrated in the last decade of the current century on developing models that assess the competitiveness of the tourism destination using a particular empirical technique.

The aforementioned scientific interest has been sparked, principally, for three reasons: the growing competitiveness amongst tourism sites, despite the fact that they haven't altered much over time; first, the growing importance of the tourism sector in the local, national, and worldwide economies; And finally, while the short-term economic benefits of tourism appear to be evident, the long-term consequences do not seem to be as positive, or at least there is not enough scientific data to support them.

8.6.2. The Conceptual Framework of Tourism Competitiveness

The conceptual framework of tourist competitiveness is founded on ground-breaking studies, among others, that started to be published at the close of the twentieth century and the beginning of the twenty-first. As a destination's capacity to draw tourists grows and overall tourist expenditure rises as a result of providing visitors with gratifying, satisfying experiences, this is what is meant when a place is said to be more competitive than another.

Additionally, according to these authors, the quality of life for locals must be improved, and continued increase of the area's natural resources is required to ensure that it can support future generations. The measurement of tourism competitiveness is a challenging task, as, in agreement with, it is an abstract concept, not directly observable, and with a multidimensional character resulting from the sum of variables, many of which are compared to those of another tourist destination that is taken as a reference. This is true even though it may be straightforward to determine the level of attraction of a geographic location by exploring the registry of data.

These models share the fact that they focus on a variety of interconnected factors that affect tourist competitiveness, recognizing that this phenomenon is undoubtedly complex and can be defined by both objective and subjective indicators.

8.6.2.1. Porter's Model

Porter's model of competitive advantage, popularly known as "Porter's Diamond," offers a mechanism for formulating business strategies to increase competitiveness in a globalized economy. According to this concept, a country's companies' competitiveness is influenced by both natural and advantageous variables that may already exist in the country as well as elements that may be produced as a result of positive or negative synergy with the ultimate aim of increasing productivity (Figure 8.5).

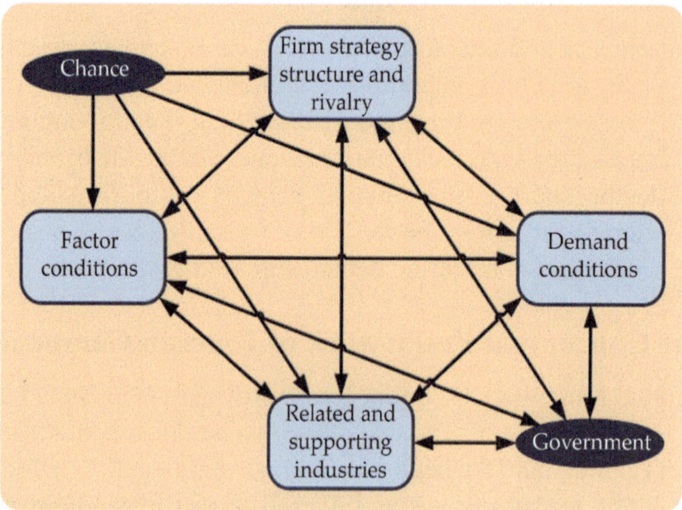

Figure 8.5. Porter's Diamond model.

Source: Image by Wikimedia Commons

8.6.2.2. The Crouch-Ritchie Model

The Crouch-Ritchie competitiveness model was the first of the models presented to evaluate the tourism competitiveness of the 'long-haul' destinations. It can be considered, in agreement with, to be the main reference model on tourism competitiveness made to date.

This claim is based on the fact that the model includes all the important factors that may determine the competitiveness of a tourism destination. it is considered that this model is neither predictive nor causal, but simply a conceptual model, as its intention is nothing more than to explain the determining factors of tourism competitiveness, by using abstract concepts and relationships.

8.6.2.3. Kim's Model

The third of the basic models of tourism competitiveness is that of Kim, which considers four sources of tourism competitiveness:

The economic agents that are involved in the tourism process, the environment and tourism resources which make the place attractive to the tourist. The tourism policies: the planning, management, investments made in the sector, taxes applied to tourism activities and the prices of the services of the sector or the prices of other sectors directly related to it.

The tourism infrastructure, the system of accommodation of the visitors, the publicity and the qualification of the human resources. The tourism demand itself, the employment generated by the sector, the tourism behavior and the tourism export, which are the result of the three first sources of competitiveness.

8.6.3. Tourism Competitiveness and Economic Growth

As was already said, several institutions on a global scale have emphasized the significance of tourism as a source of growth and development. But this visibility hasn't stopped there; there is now a significant scholarly interest in this topic emerging in the academic world.

Thus, some economists, including those who have emerged more recently, contend that the tourism industry can aid in the growth and development of nations, particularly given the ability of international travel to generate foreign exchange that can be used to finance capital investments that enhance the production of goods and services. In that sense, in 2013, the international income through tourism represented 30% of global export services, and 6% of the total global exports of goods and services. For other economists, however, tourism activity can generate growth simply because it is a "large industry." It is more — it can be considered the largest industry at global level .

As such, in agreement with, in 2013, the total contribution of tourism to the global economy reached 9.5% of global GDP (7 trillion US$), generating a total of almost 22 million jobs, which represents one in ten jobs on the planet, of which 9 million jobs, 4.1% of the total global employment, can be considered jobs directly related to tourism.

As a result, the industry, according to several authors, has a significant ability to boost employment, the value of its output, and tax income. These tax revenues can support economic development plans by encouraging increased rivalry among local businesses and those of other tourist locations, which will encourage the utilization of economies of scale.

The importance of the tourist industry as a force for economic growth has been acknowledged in numerous studies. However, the majority of these studies do not examine whether this expansion ultimately improves either economic development or population well-being. While acknowledging that increased tourism production aids in development, at least up to a certain level of wealth that only a small number of countries would have reached without tourism, they do believe that the growth of well-being is noticeably inferior to the rise of the GDP. Therefore, a large and essential topic for research and investigation is implied by these.

According to other, more recent studies, the expansion of particular economic sectors is correlated with tourism's competitiveness. A discussion over how to measure this competitiveness has been sparked by fresh research, which has led to new findings. In addition to these metrics, a number of recent studies have focused on the importance of certain variables, typically associated with economic expansion, as determinants of increased tourist competitiveness and, consequently, of the development or expansion of the tourism sector.

8.6.4. Tourism Activity and Economic Growth

Since the analysis of the concept known as tourist-led growth (TLG), which states that tourism causes economic growth, studies have multiplied that, while they all agree that there is a link between tourism activity and economic growth, they clearly differ in the magnitude of this link. The scale appears to be tied to the methodology employed, the study's temporal and geographic framework, and the method utilized to measure each case's tourism activity. The majority of the research were found to be based on the use of time series, while some were found to be based on a cross section analysis, and some studies used panel approaches, such as those used in the studies of sex differences in income.

The aforementioned investigation revealed that 87 papers had been published up until the beginning of 2013 that experimentally analyzed the link between tourism activities and economic growth. Of those, 55 demonstrated the presence of a causal relationship connecting tourism activities to economic growth, hence validating or supporting the TLG hypothesis.

From the other published research, 16 found a two-way link between the variables, 9 discovered a causal association between economic growth and tourism activity, and 4 did not discover any relationship at all between the two variables.

However, the authors note that while the majority of these research do demonstrate a connection between tourism activity and economic growth, the strength of this connection varies significantly across the various studies. This size variance appears to be related to the methodology employed, the time and geographic contexts mentioned in each study, and the method utilized to measure the tourism activity in each case.

Similar to this, subsequent reviews of the methodological approaches taken in the research that analyze the TLG hypothesis have revealed variations in the strength of this association depending on the methodology employed as well as the variable used to quantify tourism activities.

The meta-analysis performed in has supported these findings. Significant differences are shown with regard to the short or long term, the employment of extra economic growth variables in the model's explanation, and the utilization of various approaches to gauge tourism activity.

8.7. INTERACTION BETWEEN CULTURAL/ CREATIVE TOURISM AND TOURISM/CULTURAL HERITAGE INDUSTRIES

However, what is even more significant are national identities and cultural differences, which are essential to stand out from the crowd by attracting the attention of other cultures, attracting investments and developing the economy, strengthening culture and national ideas. In the twenty-first century modern global society, it is especially important to discuss the change of information, rapid pace of life, the ever-changing generations of people with the features particular to them. The World Tourism Organization and UNESCO's data indicate that the tourism industry, particularly its alternative branches, such as cultural and creative tourism, should be growing quickly.

A sizable tourism sector was created globally in the second half of the 20th century. The cultural heritage industry runs concurrently with this one. They both use the same resources, such as natural and cultural heritage, but they don't always work together, despite the fact that they ought to.

8.7.1. The Concept of Cultural Tourism

Cultural tourism is typically categorized in line with the driving forces behind the dominant travel motif. According to the literature, the following six categories of tourism are distinguished based on the destination of a trip: Business travel, professional travel, leisure travel, cognitive travel, sports travel, rural travel, and wellness travel.

Because the desire to study, "test out," other authentic cultures, examine artifacts of extinct societies, and the aim to become familiar with "other" cultures are an especially significant motif, cultural tourism is particularly closely tied to special demands of tourists.

This motivates those who work in heritage tourism and heritage to build a comprehensive perspective on cultural artifacts: an object cannot function without the setting it is intended to be used in, as they together produce a cultural landscape, such as the urbanist landscape of Vilnius Old Town (Lithuania).

It was inscribed onto the UNESCO list of world treasures in 2009. The Kernav archaeological site in Lithuania is a distinctive territorial complex of historical and archaeological significance (194.4 hectares), which was added to the UNESCO world heritage lists in 2004.

In order to conserve cultural and historical legacy, cultural tourism is crucial since it attracts more funding the more tourists who are interested in it. Cultural tourism is still crucial for promoting a nation or place on a worldwide scale.

Since the Great Travels of the sixteenth century, the primary objective of every journey has always been culture. The idea that culture was no longer the purpose of tourism first emerged in the 20th century. However, on a larger scale, both heritage tourism and the entire tourism industry, cultural values play a very important role. Over time, the various forms of tourism grew increasingly specialized and focused on a particular region. An alternative to traditional mass tourism, which Hall defines as a brief trip to and from a location where the traveler does not reside, is tourism that serves the cultural, educational, and other needs of the traveler.

Cultural tourism, in McKercher's opinion, is unquestionably the oldest of the "most modern" forms of tourism. People have traveled for ostensibly cultural purposes since Roman times; yet, until recently, they were not considered as special travelers. Travels on pilgrimages, excursions to famous sites, unique occasions, celebrations, and rituals have all been documented for a very long time. Alternative tourism is the name given to it today (Figure 8.6).

Figure 8.6. Kernavė Archaeological Site (Cultural Reserve of Kernavė) (Lithuania).

Source: Image by Wikimedia Commons

The definition of alternative tourism by Newsome, Moore, and Dowling is "responsible or specific interest-related tourism that includes small numbers of tourists under authentic, natural, or cultural conditions." This definition includes natural tourism (i.e., travel to a natural setting), cultural tourism (i.e., traveling to see one's heritage or religion), event tourism (i.e., traveling to events like sporting events or festivals), and other types of travel (i.e., educational, farm tourism and so on).

Cultural tourism becomes one of the primary areas of academic study and a hot topic of conversation around the turn of the twenty-first century.

The idea of "new tourism" can be explained by using one of Craik's quotes regarding more contemporary, flexible types of travel that are frequently accompanied by the adjectives "communication-enriching," "satisfying," "adventurous," and "directed toward individual requirements."

The fulfillment of cultural requirements becomes not just an element of vacations but increasingly the primary motivator of travel. Due to various conflicts across the world and xenophobia based on misperceptions, cultural tourism can frequently aid in cross-cultural communication and promote understanding. By learning about the cultures of different countries, we can better understand one another and feel a growing desire to work together. This encourages interaction and integration.

8.7.2. Benefit of Cultural Tourism

A sense of discovery, intercultural understanding, and personality enrichment are all provided by cultural tourism. Understanding cultural variety, traditions, customs, learning about various lifestyles, etc. are all made easier by cultural tourism. Simple human well-being is comprised of these elements: feeling, interacting, tasting, discovering, traveling, meeting others, taking part in activities, and learning.

In an effort to protect cultural treasures, UNESCO supports cultural tourism. Cultural tourism can often facilitate communication and promote understanding among groups of people who are estranged by mistrust and xenophobia in a world of war.

Thus, cultural tourism enhances intercultural communication. An individual's perception is enriched by diverse experience, which also improves their capacity to comprehend and empathize with others. People benefit from one another's knowledge and subsequently develop a variety of talents to address their own difficulties. As a result, communication and integration are simultaneously encouraged.

Tourists are more likely to spend money in areas where there are more jobs available to inhabitants, making it simpler to keep people in the area, which benefits both the national economy and the region's development. As a result of giving kids greater options, the demographic situation is improved. This kind of tourism offers numerous potential for improving living circumstances for both travelers and the hosts of a country, and its operations cover the entire target territory. Additionally, cultural tourism fosters innovation and promotes collaboration among locals, which promotes a nation's economic and social development.

8.7.3. Creative Tourism

Cultural tourism gave way to creative tourism as people's needs and desires to experience something new changed.

A novel style of travel known as "creative tourism" has the potential to drastically alter the way that travel is now done and to give rise to entirely new kinds of travel experiences. The growth of the local economy and promotion of social and cultural development may be aided by creative tourism. This kind of tourism involves not only foreign visitors who observe a country's culture first-hand but also the society that develops and upholds its traditions and traditional values.

A significant amount of influence was produced by the existence of the consumerist society with relation to these elements that have contributed to the growth of creative tourism. People were compelled to consider the past and how it needed to be preserved. The emergence of creative tourism was mostly impacted by changing thinking and having more free time, which encouraged people to seek out discoveries that were distinct from those made by mass travelers.

8.7.4. Heritage Tourism

A lot of time has passed since the idea of cultural heritage first emerged. It is possible to distinguish between the notions of monumental and resource culture legacy throughout history. The Renaissance conceptions of monumento (En. monument) and antichitá are where the idea of monumental legacy first emerged (En. antique).

The development of the international community, international law, legislative reforms, the industrialization of society, and historical events of the 20th century that demonstrated the fragility and irreversibility of cultural heritage have all had an impact on this concept's evolution since that time.

A few decades ago, the second idea of cultural heritage emerged. This idea contends that legacy shouldn't be memorialized with signs since it has a different purpose, namely serving as the foundation upon which modern culture, an integral component of human lifestyle and surroundings, grows and develops. This notion of heritage is defined in terms of resources. This idea contends that if heritage is constituted by historical resources, then such resources are finite and irreversible, necessitating their preservation and transfer to subsequent generations. Various legal acts where the definition and explanation of heritage were specified and changed determined the shift in how heritage was perceived.

In recent times, legacy is increasingly being viewed as a process that items go through or as a tool for marketing. As a result, a lot depends on how people feel and how they perceive things. The majority of important items to include in lists of legacy are those that people believe to be worth.

There are seven major categories of heritage listed by UNESCO: nature, landscape, monuments, artifacts, activity, people, and locations. Due to the fact that many cultural heritage items could fit into multiple categories, none of these categories are fully defined. One of them is cultural heritage. The Charter for the Conservation of Places of Cultural Heritage of New Zealand defines cultural heritage as "regions, landscape and features, buildings, constructions and gardens, archaeological and traditional locations, sacred places and monuments that have long-term value and can be legally assessed, that teach people about the past and the culture of those who lived there, that give context to the identity of the society as a whole, and that provide context for the identity of the people who relate to them."

8.7.5. Significance of the Interaction between Cultural Tourism and Cultural Heritage

One the one hand, historical context and significance are provided by cultural tourism for the already existent heritage institutions. However, it poses a threat to their own survival. Finding a balance between heritage and cultural tourism is actually the main challenge.

Most frequently, those with expertise in managing cultural tourism and cultural heritage work separately. Even when there are shared interests, their communication does not seem to be happening. They collaborate insufficiently, lacking the necessary teamwork to create truly excellent items.

According to the world tourist organization, experts in cultural tourism and heritage might come to a mutually beneficial accord. Cultural tourism would then become fully meaningful since it would arouse a sense of discovery, promote intercultural understanding, and start to develop a person's spiritual richness. The heritage items would then receive the funding they require for repair and become accessible to tourists. One of the key causes for protecting heritage is the growth of cultural tourism. According to the facts, if historical sites are not open to tourists and the general public, they would eventually be destroyed (such as Lithuanian mansions or parks). Both people and various natural occurrences do damage to them.

Heritage that does not yield financial gain is not preserved and repaired. In this situation, the local community stops respecting the cultural values and symbols present there. On the other hand, the historical artifacts used in cultural tourism are preserved and maintained. The guardians of heritage make sure that the artifacts are presented to society in the right way.

In order to maintain and repair them, funds from cultural tourism are needed. However, the fundamental advantage of cultural tourism is that it provides a chance for visitors to learn about the history of a nation and/or a specific location that they would most likely never encounter otherwise. The perception of a region, a state, or a nation is created in this way.

According to R. Dapkus, cultural legacy is one of the crucial components of a nations or region's development and image. The protection of cultural traditions is said to depend in part on cultural tourism. More than that, cultural tourism might raise people's standards of living.

It should be emphasized that access to very significant natural, historical, and cultural sites is rigorously regulated: different limitations have been established, and locality management plans have been worked out.

8.8. CONCLUSION

This chapter summarizes Community participation in tourism development. This chapter also explains the types of community participation, factors influencing community participation, etc. This chapter also covers Community Development Strategies for Tourism Development in Langkawi Islands, Malaysia.

The mid of this chapter provides highlights on ecotourism and its role in sustainable development, ecotourism activities that are not performed according to the purpose, principles, and characteristics cause disturbance in environmental, economic, and socio-cultural fields due to over-intensification to occur especially in sensitive ecosystems like natural and cultural areas.

Therefore, to provide sustainability in ecotourism, it is necessary to know the environmental, social, and economic effects of ecotourism activities and consider these effects during the planning.

Toward the end of the chapter, the Interaction between Cultural/Creative Tourism and Tourism/ Cultural Heritage Industries with the benefit of Cultural Tourism has been discussed.

REFERENCES

1. K. C. A., (2016). Ecotourism and its role in sustainable development of Nepal. *Tourism – From Empirical Research Towards Practical Application* [Preprint]. Available at: https://doi.org/10.5772/62308.
2. Marzuki, A., & Khoo, J., (2016). Community development strategies for tourism development in Langkawi Islands, Malaysia. *Tourism – From Empirical Research Towards Practical Application* [Preprint]. Available at: https://doi.org/10.5772/62575.
3. Pablo-Romero, M. D., Gómez-Calero, P., & Sánchez-Rivas, J., (2016). Tourism, competitiveness and economic growth: A new analytical model. *Tourism – From Empirical Research Towards Practical Application* [Preprint]. Available at: https://doi.org/10.5772/62276.
4. Rasoolimanesh, S. M., & Jaafar, M., (2016). Community participation toward tourism development and conservation program in rural World Heritage Sites. *Community Participation toward Tourism Development and Conservation Program in Rural World Heritage Sites* [Preprint]. Available at: https://doi.org/10.5772/62293.
5. Tourism – from empirical research towards practical application, (2016). *Interaction Between Cultural/Creative Tourism and Tourism/ Cultural Heritage Industries* [Preprint]. Available at: https://doi.org/10.5772/61385.

INDEX

A

Accommodations 47
Accounting officers 33
Adequate coordination 37
Adequately trained 2
Administrative mistakes 128
Administrators 70
Advanced Trauma Life Support (ATLS) 178
Adventure tourism 17
Aggregation model 159
Agricultural revenue 214
Agri-tourism 18
Amplified 205
Analytical 76
Analyzing 118
Ancient civilization 40
Ancient civilizations 40
Anthropology 44
Antibiotic stewardship 175
Archaeological tourism 15
Artifacts 44
Artificial intelligence (AI) 13
Asymmetrical cryptosystem 183
Attributes 51

B

Beach operators 161
Beverages 122
Biodiversity 212
Biodiversity preservation 213
Biological variety 213
Bio political dilemmas 173
Bitcoin anonymizers 184
Bookstore tourism 19
Boost national wealth 34

C

Cardiovascular screening 177

Casual employment 119
Catering 122
Cathedrals 145
Childcare 33
Chronic 175
Chronic health disorders (CHCs) 175
Citizen power 205
Citizens 202
Civilizational component 46
Client 72
Climate change 173
Cognitive 31
Cognitive competencies 63
Cognitive demand 150
Collaboration 25, 130
Comfort zone 117
Commodification 92
Communicate 126, 211
Communications 94
Community 155, 202
Community adapts 205
Community-based ecotourism 216
Community involvement 177
Community leadership 209
Community levels 209
Competency 117
Competitive rates 34
Conceiving 46
Conceptual gap 31
Conservation 216
Conservation programs 206
Constructivism theories 25
Consumer service 214
Consumer's perception 151
Contrasting 76
Conventional methods 68
Conventional training 111
Coordination 174
Corporate downsizing 73
Cosmetic operations 162
Counselors 8
Coworkers 127
Cruise lines 101
Cruises 122
Cryptography 182
Cultural eroding 208
Cultural heritage tourism 15
Culturally 30
Cultural tourism 140, 144
Culture 145
Customers 65

D

De-layering 73
Destination marketing 131
Destinations 13
Digitalization 24
Digital marketing 13
Digital revolution 123
Disaster tourism 19
Disciplinary interests 70
Discretionary activity 148
Distinctiveness 103
Diverse forests 215
Diversification 94, 121
Diversity 15
Dynamic interaction patterns 166
Dynamics 140

E

Eco-label tool's 161
Economic growth 3, 202
Economic influence 122
Economic sectors 100
Ecosystem tourism 213
Ecotourism 4, 211, 212, 214, 216
Ecotourism initiatives 215

Ecotourism profits 212
Editors 173
Educators 95
Egyptians 39
Emerging infectious diseases (EID) 172
Emotional intelligence 24
Empathy 122
Emphasis 70
Emphasize 140
Employability 110
Employability Skills 110
Employers 5
Employment contract 116
Encouragement 37
Encourages 206
End-of-life (EOL) 177
Energize readers 173
Entrepreneurial behaviors 62
Entrepreneurial intention 79
Entrepreneurial mindsets 65
Entrepreneurial talents 83
Entrepreneurial vocation 79
Entrepreneurs 64, 66
Entrepreneurship 9, 58
Entrepreneurship courses 66
Entrepreneurship degrees 74
Entrepreneurship education 58
Environmental sustainability 91
Esthetic surgeries 20
Event tourism 140
Evolution 81
Excursion 19
Exoticness 155
Exotic travel experiences 93
Expectancy theory of motivation (ETM) 78
Exploitation 132, 216
Extracurricular activities 76

F

Fairy-tale 157
Farming 214
Farm owner 156
Favorable cognitive 151
Favorable evaluation 147
Finance 21
Financial gains 216
Firms 130
Fluctuating population 52
Foreigners 3
Forging bonds 128
Fostering leadership 65
Full-service hotels 104

G

Generalization 53
Generic systems 50
Geographically 146
Geographic area 161
Geopolitical environment 158
Geo-tourism 21
Global economic activity 2
Global economy 88
Global health security 173
Global industry 214
Globalization 88
Globalized economy 220
Global marketization 80
Governmental domains 62
Governmental organizations 210
Graduate employment 113
Growing trend 158

H

Halal tourism 141
Harmonious blending 37
Health security 172
Health tourism 4

Hearing aids 177
Hedonic advantage 149
Hedonic benefit 147
Hiking 216
Historical landscapes 156
Historical variances 156
HIV 176
Holistic view 26
Homogenization 92
Hospitality 2
Hospitality Candidates 125
Hospitality education 2
Hospitality jobs 5
Hospitality program 74
Hospitality sector 101
Hospitality skills 13
Hotel management 83
Hotels & Resorts 2
HR 21
HRD plans 106
Human capital 98
Human phenomenon 41
Human resources managers 125
Human trafficking 172
Hypotheses 43, 45

I

IHS experts 172
IHS risks 174
Illnesses 174
Incorporating entrepreneurialism 72
Incorporating experiential learning 101
Indicators 202
Indigenous groups 215
Individualization 136
Infectious disease 172
Infectious illnesses 174
Influencing factors 140
Infrastructure 3
Innovation 65
Innovation-driven emphasis 176
Innovative 83
Inscription 207
Institutionalizing 75
Instructional strategies 67
Instructors 69
Insufficient reaction 175
Integration 226
Intellectual pillars 31
Intellectual synergy 25
Intensity 161
Interchangeably 58
Intergovernmental cooperation 159
International health security (IHS) 172
International Labor Organization (ILO) 128
International market 3
International relations 47
International tourism 214
International travelers 93
Internship experience 8
Interpersonal communication 114
Interpersonal relationships 52
Interrelated structures 42
Intra-commerce 47
Intrinsic 149
Islamic tourism 141

L

Labor-intensive 120
Labor market 82, 129
Langkawi Development Authority 208
Language proficiency 127
Leadership 102
Leakage 94

Leisure industries 135
Lifelong process 115
Limited-service hotels 104
Listening skills 25
Livestock 211
LMIR population 177
Local laborers 209
Lodging 54
Long-haul 153
Long-term career 127
Long-term sustainability 214
Low- and middle-income countries (LMIRs) 174

M

Macroeconomic conditions 166
Managerial responsibilities 119
Managers 32
Manipulative participation 205
Marine areas 161
Marketers 140
Marketing 21
Marketing firms 10
Medical tourism 20
Mediterranean 158
Mentorship 73
Methodology 160
Meticulous regulation 33
Microfluidic lab 172
Minority cultures 214
Missile silos 18
Mobilities 140
Mobility aids 177
Mobility issues 176
Mobilize the people 60
Mobilizing money 36
Monasteries 145
Monuments 44
Morbidity 177
Mortality 172, 178
Motivation 38
Multi-destination excursions 153
Multidimensional character 220
Multidisciplinary approach 77
Multinational companies (MNC) 115
Multinational corporation 116

N

Natural heritage 91
Natural resources 211, 212
Natural sceneries 156
Nearby sightseeing 155
NGOs 210
Night Clubs & Bars 2
Non-cognitive 63
Non-communicable diseases 175
Non-hospitality sectors 101
Non-vocational positions 111
Nostalgia 153
Novelty 153

O

Ongoing expansion 31
Open-air museum 14
Organization 2
Organizational 209
Organizational objectives 35
Organizing 32
Originator 48
Owners 75

P

Pandemics 175
Partnerships 131
Passengers 158

Pathogen 174
Patient 177
Pedagogies 68
Personnel 117
Personnel attrition 36
Pervasive poverty 132
Pharmaceutical 172
Philosophically 30
Philosophical structures 31
Physical resources 32
Pilgrimage 16
Placement 24
Planning gatherings 156
PMS software 102
Point-of-care (POC) technology 177
Policy makers 140
Popularity 42
Population group 175
Postmodernism 44
Potential EIDs 174
Poverty 96, 175
Power holders 206
Prehospital 182
Prestige 153
Prestigious organization 116
Preventative medication 162
Primary resource 214
Prioritization skills 126
Prior notoriety 161
Privatization 80
Proactive 77
Procedures 115, 116, 162
Productivity 34
Products 208
Professional abilities 82
Professional approach 30
Professionalism 120, 130
Progressive control 159
Psychological demands 149
Psychological underpinnings 77
psychology 44
Psycho-sensory experiences 147

Q

Qualified faculty 95

R

Ramifications 182
Ransom payments 184
Ransomware 182
Recognize 75
Reconsideration 13
Recreation 46
Recreational 46
Reducing poverty 212
Reflection 112
Researchers 79
Resort operators 208
Restaurants & Catering 2
Restrictions 83
Retail store 102
Robust hospitality 22
Rural lifestyles 207

S

Scenario 159
Scientific literature 172
Search engine optimization (SEO) 13
Seasonal employment 117
Select-service hotel 104
Self-directed learners 27
Self-directing 26
Self-discovery 73
Self-supporting 26
Sensory stimuli 147
Service-value proposition 72

Severe Acquired Respiratory Syndrome (SARS) 90
Sewage 212
Shrines 48
Sightseeing 14
Sizable market 83
Small- to medium-sized businesses (SME) 115
Social contexts 113
Social interactions 88
Socializing 177
Societal 113
Socioeconomic change 65
Socioeconomic situation 89
sociology 44
Soft skills 110
Solitary working hours 120
Sophisticated 27
Souvenirs 52
Space tourism 21
Spatial range 140
Specialized skills 114
Spending 121
Spiritually 30
Spontaneous participation 206
Stabilization 178
Staff members 130
Stakeholders 100, 175
Strict quarantine 89
Stringent requirements 116
Stroke 176
Subdomains 172
Supervisors 117
Surveys 130
Symbolic benefit 148
Symbolic gain 152
Synchronize 37

T

Technical education 100
Telecommunication 119
Telecom sector 101
Theories of motivation 30
Therapeutic 46
Thermodynamics 49
Tier activity 32
Time-consuming 7
Tourism businesses 216
Tourism decision-making 209
Tourism destinations 151
Tourism development procedures 206
Tourism industry 63, 96, 159
Tourist attractions 119
Tourist destinations 48
Tourist-led growth (TLG) 222
Tourist organization 228
Tourists 144
Tourist's cognitive perception 151
Traditional curriculum system 82
Traditional economic pillars 211
Traffic congestion 207
Training and Intranets 26
Transferrable abilities 110
Transit 90
Transportation 54, 119
Traumatic injury 177
Travel behavior 140
Trivialization 92
Typology 206

U

Unanticipated events 90
Unemployment rate 133
UNESCO 204
Unpredictable environment 60
Utilization 223

Utilizing talents 113

V

Vaccinations 134
Vector control 175
Violence 132
Visitors 3, 121
Visitor satisfaction 149
Vocational 83
Vocational education 82
Vocational education and training (VET) 111
Vocational skills 111

W

Wages 34
Webification of education 24
Web users 161
Welfare programs 33
Wellness declarations 134
Wellness tourism 4
WHS conservation 203, 205
WHS destination 204
WHS locations 206
World Tourism Organization (WTO) 144